Statistics and Computing

Series Editors:
J. Chambers
W. Eddy
W. Härdle
S. Sheather
L. Tierney

Springer
New York
Berlin
Heidelberg
Barcelona
Budapest
Hong Kong
London
Milan
Paris
Singapore
Tokyo

Statistics and Computing

James E. Gentle

Random Number Generation and Monte Carlo Methods

With 30 Illustrations

 Springer

James E. Gentle
Institute for Computational
  Sciences and Informatics
George Mason University
Fairfax, VA 22030-4444
USA

Series Editors:

J. Chambers
Bell Labs, Lucent
  Technologies
600 Mountain Ave.
Murray Hill, NJ 07974
USA

W. Eddy
Department of Statistics
Carnegie Mellon University
Pittsburgh, PA 15213
USA

W. Härdle
Institut für Statistik und
  Ökonometrie
Humboldt-Universität zu Berlin
Spandauer Str. 1
D-10178 Berlin
Germany

S. Sheather
Australian Graduate School
  of Medicine
PO Box 1
Kensington
New South Wales 2033
Australia

L. Tierney
School of Statistics
University of Minnesota
Vincent Hall
Minneapolis, MN 55455
USA

Library of Congress Cataloging-in-Publication Data
Gentle, James E., 1943–
    Random number generation and monte carlo methods / James E.
  Gentle.
        p.   cm. – (Statistics and computing)
    Includes bibliographical references and indexes.
    ISBN 0-387-98522-0 (hardcover : alk. paper)
    1. Monte Carlo method.   2. Random number generators.   I. Title.
  II. Series.
  QA298.G46   1998
  519.2' 82–dc21   98-16709

Printed on acid-free paper.

Production managed by Anthony K. Guardiola; manufacturing supervised by Thomas King.
Photocomposed pages prepared from the author's LaTeX files.
Printed and bound by Maple-Vail Book Manufacturing Group, York, PA.
Printed in the United States of America.

9 8 7 6 5 4 3 2 1

ISBN 0-387-98522-0 Springer-Verlag New York Berlin Heidelberg   SPIN 10677281

To María

Preface

The role of Monte Carlo methods and simulation in all of the sciences has increased in importance during the past several years. These methods are at the heart of the rapidly developing subdisciplines of computational physics, computational chemistry, and the other computational sciences. The growing power of computers and the evolving simulation methodology have led to the recognition of computation as a third approach for advancing the natural sciences, together with theory and traditional experimentation. Monte Carlo is also a fundamental tool of computational statistics. At the kernel of a Monte Carlo or simulation method is random number generation.

Generation of random numbers is also at the heart of many standard statistical methods. The random sampling required in most analyses is usually done by the computer.

The computations required in Bayesian analysis have become viable because of Monte Carlo methods. This has led to much wider applications of Bayesian statistics, which, in turn, has led to development of new Monte Carlo methods and to refinement of existing procedures for random number generation.

Various methods for generation of random numbers have been used. Sometimes processes that are considered random are used, but for Monte Carlo methods, which depend on millions of random numbers, a physical process as a source of random numbers is generally cumbersome. Instead of "random" numbers, most applications use "pseudorandom" numbers, which are deterministic but "look like" they were generated randomly. Chapter 1 discusses methods for generation of sequences of pseudorandom numbers that simulate a uniform distribution over the unit interval $(0, 1)$. These are the basic sequences from which are derived pseudorandom numbers from other distributions, pseudorandom samples, and pseudostochastic processes.

In Chapter 1, as elsewhere in this book, the emphasis is on methods that *work*. Development of these methods often requires close attention to details. For example, whereas many texts on random number generation use the fact that the uniform distribution over $(0, 1)$ is the same as the uniform distribution over $(0, 1]$ or $[0, 1]$, I emphasize the fact that we are simulating this distribution with a discrete set of "computer numbers". In this case whether 0 and/or 1 is included does make a difference. A uniform random number generator should not yield a 0 or 1. Many authors ignore this fact. I learned it almost twenty years ago shortly after beginning to design industrial-strength software.

Chapter 2 discusses general methods for transforming a uniform random deviate or a sequence of uniform random deviates into a deviate from a different distribution. One or more of these methods may be used to generate random deviates from a distribution that is not one of the standard distributions.

Chapter 3 describes methods for some common specific distributions. It is not the purpose of Chapter 3 just to catalog a large number of methods. In some cases, only the best method is described; in other cases, especially where the best method is very complicated, a simple method is described and references are given for other methods. Although some of the exercises require the reader to write a program to generate random numbers, for production work I recommend using existing software whenever it is available.

Chapters 1, 2, and 3 are the longest chapters in the book, covering most of the basic methods. Chapter 4 continues the developments of Chapters 2 and 3 to apply them to generation of samples and of nonindependent sequences.

Chapter 5 considers some applications of random numbers. Some of these applications are to solve deterministic problems. This is called Monte Carlo.

The Monte Carlo methods raise questions about the quality of the pseudorandom numbers that simulate physical processes and about the ability of those numbers to cover the range of a random variable adequately. In Chapter 6, I address some of these issues, including the basic question of whether we should even attempt to simulate a random sequence. In many cases a quasirandom sequence that has regular patterns may be more appropriate.

Chapter 7 provides information on computer software for generation of random variates. The discussion concentrates on two software packages, IMSL and S-Plus.

Monte Carlo methods are widely used in the research literature to evaluate properties of statistical methods. Chapter 8 addresses some of the considerations that apply to this kind of study. It is emphasized that the study uses an *experiment*, and the principles of scientific experimentation should be observed.

The literature on the topic of random number generation and Monte Carlo methods is vast. The reader may rightly infer from some of the subsequent discussion that I believe much of the literature to be barely incremental or even redundant. Although the list of references beginning on page 211 is rather extensive, I do not attempt to sort out the literature; and I do not attempt to present a comprehensive bibliography. For bibliographies of early work in this area the reader is referred to Sowey (1972, 1978, 1986). In many cases where I do not discuss a topic, I simply cite a reference in an attempt at completeness. (Although it could be useful to list references that make no real contribution and to state this fact when the reference is cited, I have generally omitted references that do not add anything to the field. Because of not keeping track of these, however, in my own studies I have often wasted time by tracking down a reference for the second or third time just to find nothing of substance.)

This book has grown from lecture notes I have used in different courses in the computational and statistical sciences over the past few years. The book would be appropriate for a course in Monte Carlo methods or simulation. It could also be used for supplemental or reference material in a course in statistical computing, computational statistics, or other computational sciences.

The material in this book is part of the field of statistical computing, which comprises computational methods that support statistical methods. In other notes that form the larger whole of which this book is a part, I discuss techniques of computational statistics. The field of computational statistics includes the set of statistical methods that are computationally intensive and use computation as a basic tool of discovery, as opposed to computation for the purpose of processing data according to a fixed paradigm.

Many methods of computational statistics, such as cross-validation, Markov chain Monte Carlo, and bootstrap, rely on random number generation and the basic Monte Carlo technique. The purpose of the present text is to cover the basic techniques. The applications in computational statistics are referred to, but not discussed in detail.

The main prerequisite for this text is some background in what is generally called "mathematical statistics". Some scientific computer literacy is also necessary. I do not use any particular software system in the book, but I do assume the ability to program in either Fortran or C, and the availability of either S-Plus, Matlab, or Maple. For some exercises, the required software can be obtained from either statlib or netlib (see the bibliography).

When I have taught this material, my lectures have consisted in large part of working through examples. Some of those examples have become exercises in the present text. The exercises should be considered an integral part of the book.

In every class I teach in computational statistics, I give Exercise 8.3 in Chapter 8 (page 191) as a term project. It is to replicate and extend a Monte Carlo study reported in some recent journal article. Each student picks an article to use. The statistical methods studied in the article must be ones that the student understands, but that is the only requirement as to the area of statistics addressed in the article. I have varied the way the project is carried out, but it always involves more than one student working together. A simple way is for each student to referee another student's first version (due midway through the term), and to provide a report for the student author to use in a revision. Each student is both an author and a referee. In another variation, I have students be coauthors. One student selects the article, and designs and performs the Monte Carlo experiment, and another student writes the article, in which the main content is the description and analysis of the Monte Carlo experiment.

Acknowledgments

Over the years, I have benefited from associations with top-notch statisticians, numerical analysts, and computer scientists. There are far too many to acknowledge individually, but four stand out. My first real mentor — who was a giant in each of these areas — was Hoh Hartley. My education in statistical computing continued with Bill Kennedy, as I began to find my place in the broad field of statistics. During my years of producing software used by people all over the world, my collaborations with Tom Aird helped me better to understand some of the central issues of mathematical software. Finally, during the past several years, my understanding of computational statistics has been honed through my association with Ed Wegman. I thank these four people especially.

I thank my wife María, to whom this book is dedicated, for everything.

I used TEX via LATEX to write the book, and I used S-Plus and Matlab to generate the graphics. I did all of the typing, programming, etc., myself (mostly early in the morning or late at night), so all mistakes are mine.

Material relating to courses I teach in the computational sciences is available over the World Wide Web at the URL,
    http://www.science.gmu.edu/

Notes on this book, including errata, are available at
    http://www.science.gmu.edu/~jgentle/rngbk/

Notes on a larger book in computational statistics are available at
    http://www.science.gmu.edu/~jgentle/cmpstbk/

Fairfax County, Virginia   James E. Gentle
                                                               June 17, 1998

Contents

Chapter 1

Simulating Random Numbers from a Uniform Distribution

Because many statistical methods rely on random samples, applied statisticians often need a source of "random numbers". Older reference books for use in statistical applications contained tables of random numbers, which were intended to be used in selecting samples or in laying out a design for an experiment. Statisticians now rarely use printed tables of random numbers, but occasionally computer-accessed versions of such tables are used. Far more often, however, the computer is used to generate "random" numbers directly.

The use of random numbers has expanded beyond random sampling or random assignment of treatments to experimental units. More common uses now are in simulation studies of physical processes, of analytically intractable mathematical expressions, or of a population resampling from a given sample from that population. Although we do not make precise distinctions among the terms, these three general areas of application are sometimes called "simulation", "Monte Carlo", and "resampling".

Randomness and Predictability

The digital computer cannot generate random numbers, and it is generally not convenient to connect the computer to some external source of random events. For most applications in statistics, this is not a disadvantage if there is some source of *pseudorandom* numbers, samples of which *seem* to be randomly drawn from some *known* distribution. There are many methods that have been suggested for generating such pseudorandom numbers. (As a side comment here, it might be noted that there are *two* issues: randomness and knowledge of the distribution. Although, at least heuristically, there are many external processes

1

that could perhaps be used as sources of *random* numbers — rather than pseudorandom numbers there would still be the issue of *what is the distribution* of the realizations of that external random process. For random numbers to be useful, their distribution must be known.)

Deterministic generators yield numbers in a fixed sequence such that the previous k numbers (usually just the single previous number) determine(s) the next number. Therefore, because the set of numbers usable in the computer is finite, the sequence will repeat. The length of the sequence prior to beginning to repeat is called the "period", or cycle length. (Sometimes it is necessary to be more precise in defining the period to account for the facts that with some generators, different starting subsequences will yield different periods, and that the repetition may begin prior to reaching the initial state.)

Random number generation has applications in cryptography, where the requirements for "randomness" are generally much more stringent than for ordinary applications in simulation. In cryptography, the objective is somewhat different, leading to a dynamic concept of randomness that is essentially one of predictability: a process is "random" if the known conditional probability of the next event, given the previous history (or any other information, for that matter) is no different from the known unconditional probability. (The condition of being "known" in such a definition is a primitive, i.e., undefined, concept.) This kind of definition leads to the concept of a "one-way function" (see Luby, 1992). A one-way function is a function f, such that for any x in its domain, $f(x)$ can be computed in polynomial time, and given $f(x)$, x cannot be computed in polynomial time. ("Polynomial time" means that the time required can be expressed as or bounded by a polynomial in some measure of the size of the problem.) The existence of a one-way function has not been proven. In random number generation, the function of interest yields a stream of "unpredictable" numbers; that is,

$$x_i = f(x_{i-1}, x_{i-2}, \cdots, x_{i-k})$$

is easily computable; but x_{i-1}, given x_i, x_{i-2}, ..., x_{i-k}, is not easily computable. The generator of Blum, Blum, and Shub (1986) (see page 25) is unpredictable under certain assumptions. A survey of some uses of random numbers in cryptography is available in Lagarias (1993).

Boyar (1989) and Krawczyk (1992) consider the general problem of predicting the output of random number generators. They define the problem as a game in which a predictor produces a guess of the next value to be generator and the generator then provides the value. If the period of the generator is p, clearly a naive guessing scheme would become successful after p guesses. For certain kinds of common generators, Boyar (1989) and Krawczyk (1992) give methods of predicting the output in which the number of guesses can be bounded by a polynomial in $\log p$. The reader is referred to those papers for the details.

Terminology

Although we understand that the generated stream of numbers is really only pseudorandom, in this book we usually use just the term "random", except when we want to emphasize the fact that the process is not really random, and then we use the term "pseudorandom". Pseudorandom numbers are meant to simulate random sampling. Generating pseudorandom numbers is the subject of this chapter. In Section 6.3, page 159, we consider an approach that seeks to insure that, rather than appear to be a random sample, the generated numbers are more uniformly spread out over their range. Such a sequence of numbers is called a quasirandom sequence.

We use the terms "random number generation" (or "generator") and "sampling" (or "sampler") interchangeably.

Another note on terminology: Some authors distinguish "random numbers" from "random variates". In their usage, the term "random numbers" applies to pseudorandom numbers that arise from a uniform distribution, and the term "random variates" applies to pseudorandom numbers from some other distribution. Some authors use the term "random variates" only when those numbers resulted from transformations of "random numbers" from a uniform distribution. I do not understand the motivation for these distinctions, so I do not make them. In this book "random numbers" and "random variates", as well as the additional term "random deviates", are all used interchangeably. We generally use the term "random variable" with its usual meaning, which is different from the meaning of the other terms. *Random numbers* or *random variates* simulate realizations of *random variables*. I will also generally follow the notational convention of using capital Latin letters for random variables and corresponding lower case letters for their realizations.

In most cases we want the generated pseudorandom numbers to simulate a uniform distribution over the unit interval $(0, 1)$, that is, the distribution with the probability density function,

$$
\begin{aligned}
p(x) \;&=\; 1, \quad \text{if} \;\; 0 < x < 1; \\
&=\; 0, \quad \text{otherwise.}
\end{aligned}
$$

We denote this distribution by $U(0, 1)$. The uniform distribution is a convenient one to work with because there are many techniques available to transform the uniform samples into samples from other distributions of interest.

There are currently two basic techniques in common use for generating uniform random numbers: congruential methods and feedback shift register methods. In both cases, usually random integers over some fixed range are first generated and then scaled into the interval $(0, 1)$. If the range of the integers is large enough, the resulting granularity is of little consequence in modeling a continuous distribution. (The granularity of random numbers from good generators is no greater than the granularity of the numbers with which the computer ordinarily works.)

Modular Arithmetic

Both basic methods, congruential and feedback shift registers, use modular arithmetic, so we now describe a few of its properties. For more details on general properties the reader is referred to Shockley (1967), Fang and Wang (1994), or some other text on number theory. Zaremba (1972) discusses several specific applications of number theory in random number generation and other areas of numerical analysis.

The basic relation of modular arithmetic is *equivalence modulo m*, also called *congruence modulo m*, where m is some integer. Two numbers are said to be equivalent, or congruent, modulo m if their difference is an integer evenly divisible by m. For a and b, this relation is written as

$$a \equiv b \bmod m.$$

For example, 5 and 14 are congruent modulo 3 (or just "mod 3"); 5 and -1 are also congruent mod 3. Likewise 1.33 and 0.33 are congruent mod 1. It is clear from the definition that congruence is symmetric:

$$a \equiv b \bmod m \quad \text{implies} \quad b \equiv a \bmod m;$$

reflexive:
$$a \equiv a \bmod m \quad \text{for any } a;$$

and transitive:

$$a \equiv b \bmod m \quad \text{and} \quad b \equiv c \bmod m \quad \text{implies} \quad a \equiv c \bmod m;$$

that is, congruence is an "equivalence relationship".

A basic operation of modular arithmetic is *reduction modulo m*; that is, for a given number b, find a such that $a \equiv b \bmod m$ and $0 \le a < m$. If a satisfies these two conditions, a is called the *residue* of b modulo m.

Reduction of b modulo m can also be defined as

$$a = b - \lfloor b/m \rfloor m,$$

where the floor function $\lfloor \cdot \rfloor$ is the greatest integer less than or equal to the argument.

From the definition of congruence, we see that the numbers a and b are congruent modulo m if and only if there exists an integer k such that

$$km = a - b.$$

(In this expression a and b are not necessarily integers, but m and k are.) This consequence of congruence is very useful in determining equivalence relationships. For example, using this property it is easy to see that if

$$a \equiv b \bmod m$$

and
$$c \equiv d \bmod m,$$
then
$$ac \equiv bd \bmod m.$$

A system of modular arithmetic is usually defined on the nonnegative integers. Modular reduction together with the two operations of the ring results in a finite field on a set of integers. In using congruential random number generators it is common to work with a finite field of integers consisting of the nonnegative integers that are directly representable in the computer, that is, of about 2^{31} integers.

Because the pseudorandom numbers we wish to generate are between 0 and 1, in some algorithms reduction modulo 1 is used. The resultants are the fractional parts of real numbers.

Modular arithmetic has some useful applications with true random variables also. An interesting fact, for example, is that if R is a random variable distributed as $U(0,1)$ and

$$S \equiv (kR + c) \bmod 1$$

where k is an integer constant not equal to 0, and c is a real constant, then S has a $U(0,1)$ distribution. (You are asked to show this in Exercise 1.1a, page 38.) Modular arithmetic can also be used to generate two independent random numbers from a single random number. If

$$\pm 0.d_1 d_2 d_3 \cdots$$

is the representation, in a given base, of a uniform random number R, then any subsequence of the digits d_1, d_2, etc., can be used to form other uniform numbers. (If the subsequence is finite, as of course it is in computer applications, the numbers are discrete uniform; but if the subsequence is long enough the result is considered continuous uniform.) Furthermore, any two disjoint subsequences, can be used to form *independent* random numbers.

The sequence of digits $d_1, d_2, \ldots$ can be rearranged to form more than one uniform variate, for example,

$$\pm 0.d_1 d_3 d_5 \cdots$$

and

$$\pm 0.d_2 d_4 d_6 \cdots$$

The use of subsequences of bits in a fixed-point binary representation of pseudorandom numbers to form other pseudorandom numbers is called *bit stripping*.

Modular reduction is a binary operation, or a function with two arguments. In the C programming language the operation is represented as "b%m". (There is no obvious relation of the symbolic value of "%" to the modular operation. No committee passed judgment on this choice before it became a standard part of the language. Sometimes design by committee helps.) In Fortran the operation is specified by the function "`mod(b,m)`"; in Matlab, the function "`rem(b,m)`"; and in Maple, "`b mod m`". There is no modulo function in S-Plus, but the operation can be implemented using the "`floor`" function as was shown above.

Modular reduction can be performed by using the lower-order digits of the representation of a number in a given base. For example, taking the two lower-order digits of the ordinary base-ten representation of a negative integer yields the decimal representation of the number reduced modulo 100. When numbers represented in a fixed-point scheme in the computer are multiplied, except for consideration of a sign bit, the product when stored in the same fixed-point scheme is the residue of the product modulo the largest representable number. In a twos-complement representation, if the sign bit is changed, the meaning of the remaining bits is changed. For positive integers x and y represented in the fixed-point variables `ix` and `iy` in 32-bit twos-complement, the product

$$iz = ix*iy$$

contains either $xy \bmod 2^{31}$ or $xy \bmod 2^{31} - 2^{31}$, which is negative.

1.1   Linear Congruential Generators

D. H. Lehmer in 1948 (see Lehmer, 1951) proposed the *linear congruential generator* as a source of random numbers. In this generator, each single number determines its successor. The form of the generator is

$$x_i \equiv (ax_{i-1} + c) \bmod m, \quad \text{with} \quad 0 \le x_i < m; \tag{1.1}$$

a is called the "multiplier"; c is called the "increment"; and m is called the "modulus" of the generator. Often c in (1.1) is taken to be 0, and in this case, the generator is called a "multiplicative congruential generator":

$$x_i \equiv ax_{i-1} \bmod m, \quad \text{with} \quad 0 < x_i < m. \tag{1.2}$$

For $c \ne 0$, the generator is sometimes called a "mixed congruential generator". The starting value in the recursion, x_0, is called the "seed". A sequence resulting from the recursion (1.1) is called a *Lehmer sequence*. Each x_i is scaled into the unit interval (0,1) by division by m, that is,

$$u_i = x_i/m.$$

If a and m are properly chosen, the u_i's will "look like" they are randomly and uniformly distributed between 0 and 1.

The recurrence in (1.2) for the integers is equivalent to the recurrence

$$u_i \equiv au_{i-1} \bmod 1, \quad \text{with} \quad 0 < u_i < 1.$$

This recurrence has some interesting relationships to the first-order linear autoregressive model

$$U_i = \rho u_{i-1} + E_i,$$

where ρ is the autoregressive coefficient and E_i is a random variable with a $U(0,1)$ distribution (see Lawrance, 1992).

Because x_i is determined by x_{i-1} and since there are only m possible different values of the x's, the maximum period or cycle length of the linear congruential generator is m. Also, since $x_{i-1} = 0$ cannot be allowed in a multiplicative generator, the maximum period of the multiplicative congruential generator is $m - 1$.

When computing was expensive, values of m used in computer programs were often powers of 2. Such values could result in faster computer arithmetic. The maximum period of multiplicative generators with such moduli is $m/4$; and, interestingly, this period is achieved for any multiplier that is ± 3 mod 8 (see Knuth, 1981). The bits in the binary representations of the sequences from such generators have very regular patterns. The period of the lowest order bit is at most 1 (i.e., it is always the same); the period of the next lowest order bit is at most 2; the period of the next lowest order bit is at most 4; and so on. In general, low order bits in the streams resulting from a composite modulus will have periods corresponding to the factors. The small periods can render a bit stripping technique completely invalid.

Currently the numbers used as moduli in production random number generators are often primes in particular, Mersenne primes, which have the form $2^p - 1$. (For any prime $p \leq 31$, numbers of that form are prime except for the three values: $p = 11$, 23, and 29. Most larger values of p do not yield primes. A large one that does yield a prime is $p = 859\,433$.) If (and only if) m is a prime and the multiplier, a, is a primitive root modulo m, then the generator has a maximal period of $m - 1$. (A primitive root, a, modulo a prime, m, is a number such that the smallest positive k satisfying

$$a^k \equiv 1 \bmod m$$

is $m - 1$. See Shockley, 1967, for a general discussion; see Fuller, 1976, for methods to determine if a number is a primitive root; and see Exercise 1.11, page 40 for some computations.) For example, consider $m = 31$ and $a = 7$ and begin with $x_0 = 19$. The next integers in the sequence are

$$9, 1, 7, 18, 2, 14, 5, 4, 28, 10, 8, 25, 20, 16, 19,$$

so, of course, at this point the sequence begins to repeat. The period is 15; 7 is not a primitive root modulo 31. If, on the other hand, we take $a = 3$ and begin with $x_0 = 19$, we go through 31 numbers before we get back to 19; 3 is a primitive root modulo 31.

The most commonly used modulus is probably the Mersenne prime $2^{31} - 1$. A common multiplier for that modulus is 7^5 (see the discussion of the "minimal standard" on page 18). The Mersenne prime $2^{61} - 1$ is also used to some extent. Wu (1997) suggests multipliers of the form $\pm 2^{k_1} \pm 2^{k_2}$ because they result in particularly simple computations, yet seem to have good properties. Other moduli that are often used are $2^{32} - 2$ and $2^{31} - 5$.

For a random number generator to be useful in most practical applications, the period must be of the order of at least 10^9 or so, which means that the modulus in a linear congruential generator must be at least that large. The values of the moduli in common use range in order from about 10^9 to 10^{15}. Even so, the period of such generators is relatively short because of the speed with which computers can cycle through the full period and in view of the very large sizes of some simulation experiments.

1.1.1   Structure in the Generated Numbers

In addition to concern about the length of the period, there are several other considerations. It is clear that if the period is m, the output of the generator over a full cycle will be evenly distributed over the unit interval. If we ignore the sequential order of a full-period sequence from a congruential generator, it will appear to be $U(0,1)$; in fact, the sample would appear *too much* like a sample from $U(0,1)$.

A useful generator, however, must generate *subsamples* of the full cycle that appear to be uniformly distributed over the unit interval. Furthermore, the numbers should appear to be distributionally independent of each other; that is, the serial correlations should be small.

Unfortunately, the structure of a sequence resulting from a linear congruential generator is very rigid. Marsaglia (1968) pointed out that the output of any congruential generator lies on a simple lattice in a k-space with axes representing successive numbers in the output. This is fairly obvious upon inspection of the algebra of the generator. How bad this is (that is, how much this situation causes the output to appear nonrandom) depends on the structure of the lattice. A lattice is defined in terms of integer combinations of a set of "basis vectors". Given a set of linearly independent vectors $\{v_1, v_2, \ldots, v_d\}$ in $\mathbb{R}^d$, a lattice is the set of vectors w of the form $\sum_{i=1}^{d} z_i v_i$, where z_i are integers. The set of vectors $\{v_i\}$ is a basis for the lattice. Figure 1.1 shows a lattice in two-dimensions, with basis $\{v_1, v_2\}$.

For an example of the structure in a stream of pseudorandom numbers produced by a linear congruential generator, consider the output of the generator (1.2) with $m = 31$ and $a = 3$ that begins with $x_0 = 9$. The next integers in the sequence are

27, 19, 26, 16, 17, 20, 29, 25, 13, 8, 24, 10, 30, 28, 22, 4, 12, 5, 15, 14, 11, 2, 6, 18, 23, 7, 21, 1, 3, 9,

and, of course, at this point the sequence begins to repeat. The period is 30; we know 3 is a primitive root modulo 31. A visual assessment or even computation of a few descriptive statistics does not raise serious concerns about whether this represents a sample from a discrete uniform distribution over the integers from 1 to 30. The scaled numbers (the integers divided by 30) have a sample mean of 0.517 and a sample variance of 0.86. Both of these values are consistent

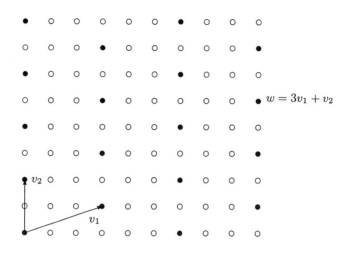

Figure 1.1: A Lattice in 2-D

with the expected values from a $U(0, 1)$ distribution. The autocorrelations for lags 1 through 5 are

$$0.27, 0.16, -0.10, 0.06, 0.07.$$

Although the lag 1 correlation is somewhat large, for a sample of this size, these values are not inconsistent with a hypothesis of independence.

However, when we plot the successive (overlapping) pairs,

$$(27, 19), (19, 26), (26, 16), \ldots$$

as in Figure 1.2, a disturbing picture emerges. All points in the lattice of pairs lie along just 3 lines, each with a slope of 3. (There are also 10 lines with slope $-\frac{1}{10}$; and 10 lines with slope $\frac{2}{11}$, if we count as lines the continuation by "wrapping" modulo 31.) Even though we have worked with overlapping pairs, it should be obvious that nonoverlapping pairs have the same structure; there are just half as many points.

This pattern is in fact related to the relatively large correlation at lag 1. Although the correlation may not appear so large for the small sample size, even if we were to increase the sample size by generating more random numbers, that large value of the correlation would persist, because the random numbers would just repeat themselves. It is easy to see that this kind of pattern results from the small value of the multiplier. The same kind of problem would also result from a multiplier that is too close to the modulus, such as $a = 27$, for example.

There are 8 primitive roots modulo 31, so we might try another one, say

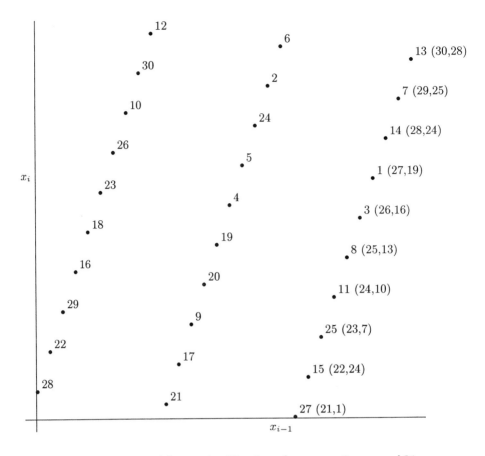

Figure 1.2: Pairs of Successive Numbers from $x_i \equiv 3x_{i-1} \bmod 31$

12. Let $a = 12$ and again begin with $x_0 = 9$. The next integers in the sequence are

15, 25, 21, 4, 17, 18, 30, 19, 11, 8, 3, 5, 29, 7, 22, 16, 6, 10, 27, 14, 13, 1, 12, 12, 20, 23, 28, 26, 2, 24, 9.

A visual assessment does not show much difference in this sequence and the sequence of numbers from the sequence generated by $a = 3$. The numbers themselves are exactly the same as those before, so the static properties of mean and variance are the same. The autocorrelations are different, however. For lags 1 through 5 they are

$$-0.01, \ -0.07, \ -0.17, \ -0.15, \ 0.03, \ 0.35.$$

The smaller value for lag 1 indicates that the structure of successive pairs may be better; and in fact, the points do appear better distributed, as we see in Figure 1.3. There are 6 lines with slope $-\frac{2}{5}$ and 7 lines with slope $\frac{5}{3}$.

A quantitative measure of the severity of the lattice structure is the distance between the lines — specifically, the shortest distance between two sides of the maximal volume parallelogram formed by four points and not enclosing any points. The distance between the lines with slope of $\frac{5}{3}$ is 6.96, as shown in Figure 1.3. The distance between the lines with slope of $-\frac{2}{5}$ is 5.76. Dieter (1975) discusses the general problem of determining the distance between the lattice lines. We encounter similar structural problems later in this section and discuss the identification of this kind of structural problem in Section 6.1, page 151.

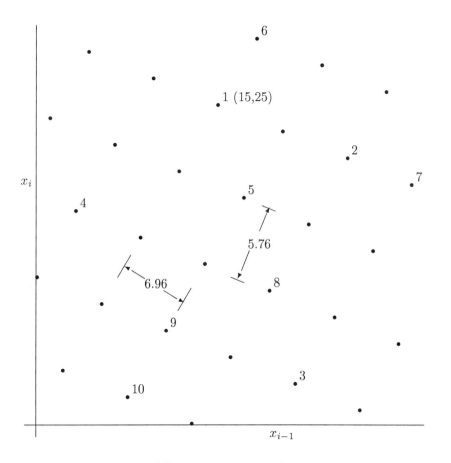

Figure 1.3: Pairs of Successive Numbers from $x_i \equiv 11x_{i-1} \bmod 31$

Figures 1.2 and 1.3 show all of the points from the full period of those small generators. For a generator with a larger period, we obviously would get more points; but with a poor generator, they could still all lie along a small number of lines.

It is a good idea to view a similar plot for a sample of points from any random number generator that we plan to use. For example, the S-Plus commands

```
xunif <- runif(1000)
plot(xunif[1:999],xunif[2:1000])
```

yield the plot shown in Figure 1.4. There does not appear to be any obvious pattern in those points generated by the standard S-Plus generator. We discuss random number generation in S-Plus in Section 7.4, page 172.

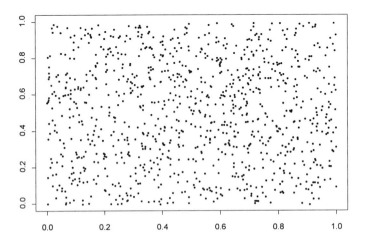

Figure 1.4: Pairs of Successive Numbers from the S-Plus Function `runif`

The two-dimensional patterns are related to the autocorrelation of lag 1, as we have seen. Autocorrelations at higher-order lags are also of concern. The lattice structure in higher dimensions is related to, but more complicated than, simple bivariate autocorrelations. In d dimensions, it is the pattern of the subsequences $(x_i, \ldots x_{i+d})$. Among triplets, for example, we could observe a three-dimensional lattice on which the points would all lie. As in the two-dimensional case, the quality of the lattice structure depends on how many lattice points are covered, and in what order they are covered. The lattice structure is related to the correlation at a lag corresponding to the dimension of the lattice; so large correlations of lag 2, for example, would suggest that a three-dimensional lattice structure would not cover three-dimensional space well.

Correlations of lag 1 in a sequence produced by a multiplicative congruential generator will be small if the square of the multiplier is approximately equal to the modulus. In this case, however, the correlations of lag 2 are likely to be large, and consequently the three-dimensional lattice structure will be poor. (In our examples in Figures 1.2 and 1.3, we had $a^2 \equiv 9 \bmod 31$ and $a^2 \equiv 28 \bmod 31$, respectively, so the generator used in Figure 1.3 would have poor lattice structure in three dimensions.) An example of a generator with very good two-dimensional lattice structure yet very poor three-dimensional lattice structure is implemented in the program RANDU, which for many years was the most widely used random number generator in the world.

The generator in RANDU is essentially

$$x_i \equiv 65539 x_{i-1} \bmod 2^{31}. \tag{1.3}$$

(RANDU was coded to take advantage of integer overflow and made an adjustment when overflow occurred, so the generator is not exactly the same as (1.3).)

The generator in (1.3) can be written to express the relationship among three successive members of the output sequence:

$$
\begin{aligned}
x_i &\equiv (65\,539)^2 x_{i-2} \bmod 2^{31} \\
&\equiv (2^{16} + 3)^2 x_{i-2} \bmod 2^{31} \\
&\equiv (6x_{i-1} - 9x_{i-2}) \bmod 2^{31}
\end{aligned}
$$

That is,

$$x_i - 6x_{i-1} + 9x_{i-2} = c2^{31}, \tag{1.4}$$

where c is an integer. Since $0 < x_i < 2^{31}$, all such triplets must lie on no more than 15 planes in $\mathbb{R}^3$. Lewis and Orav (1989) show a slice of a cube in which is plotted a sample from this generator. As might be expected, all of the points in the slice appear in only a small number of linear regions. (See Exercise 1.4, page 39.) Cabrera and Cook (1992) use data generated by RANDU to illustrate a method of projection pursuit, and their study exposes the poor structure of the output of RANDU.

Although it had been known for many years that the generator (1.3) had problems (see Coldwell, 1974), and even the analysis represented by equation (1.4) had been performed, the exact nature of the problem has sometimes been misunderstood. For example, as James (1990) states "We now know that any multiplier congruent to 5 mod 8 ... would have been better ...." Being congruent to 5 mod 8 does not solve the problem. Such multipliers have the same problem if they are close to 2^{16}, for the same reason (see Exercise 1.5, page 39).

RANDU is still available at a number of computer centers and is used in some statistical analysis and simulation packages.

The lattice structure of the common types of congruential generators can be assessed by the spectral test of Coveyou and MacPherson (1967) (see Knuth, 1981, pages 89-113) or by the lattice test of Marsaglia (1972a). We discuss these types of tests in Section 6.1, page 151.

1.1.2 Skipping Ahead in Linear Congruential Generators

Sometimes it is useful to generate separate, independent subsequences with the same generator. This may be because computations are being performed in parallel or because a Monte Carlo experiment is being run in blocks.

To generate separate subsequences, it is generally not a good idea to choose two seeds arbitrarily for the subsequences because we generally have no information about the relationships between them. In fact, an unlucky choice of seeds could result in a very large overlap of the subsequences. A better way is to fix the seed of one subsequence and then to skip a known distance ahead to start the second subsequence.

The basic equivalence relation of the generator

$$x_{i+1} \equiv ax_i \bmod m$$

implies

$$x_{i+k} \equiv a^k x_i \bmod m.$$

This provides a simple way of skipping ahead in the sequence generated by a linear congruential generator. This may be useful in parallel computations, where we may want one processor to traverse the sequence

$$x_s, \ x_{s+1}, \ x_{s+2}, \ \cdots$$

and a second processor to traverse the nonoverlapping sequence,

$$x_{s+k}, \ x_{s+k+1}, \ x_{s+k+2}, \ \cdots.$$

The seed for the second stream can be generated by

$$x_{s+k-1} \equiv bx_0 \bmod m,$$

where

$$b \equiv a^k \bmod m.$$

Another interesting subsequence that is "independent" of the first sequence is

$$x_s, \ x_{s+k}, \ x_{s+2k}, \ \cdots \qquad (1.5)$$

This sequence is generated by

$$x_{i+1} \equiv bx_i \bmod m,$$

where b is as before. This method of generating independent streams is called *leapfrogging* by k. Different "independent" subsequences can be formed using various leapfrog distances. (The distances must be chosen carefully, of course. A minimum requirement is that the distances be relatively prime.) We could let one processor leapfrog beginning at x_s and let a second processor leapfrog beginning at x_{s+1} and using the same leapfrog distance.

Sometimes, instead of "independent" sequences, we may want a sequence that is strongly negatively correlated with the first sequence. (We call such sequences "antithetic". They may be useful in variance reduction; we discuss them further in Section 5.3, page 135.) If one sequence begins with the seed x_0, and another sequence begins with $m - x_0$, then the ith term in the two sequences will be x_i and $m - x_i$. The sequences are perfectly negatively correlated.

A certain amount of care is necessary in choosing the seed and the distance to skip ahead. Anderson and Titterington (1993) show that for a multiplicative congruential generator, the correlation between the sequences $(x_i, \ldots)$ and $(x_{i+k}, \ldots)$ is approximately asymptotically $(n_1 n_2)^{-1}$, where $x_{i+k} = (n_1/n_2)x_i$ and n_1 and n_2 are relatively prime. Figure 1.5 shows a plot of two subsequences, each of length 100, whose seeds differ by a factor of 5. The correlation of these two subsequences is 0.375. (See also Exercise 1.12, page 40.)

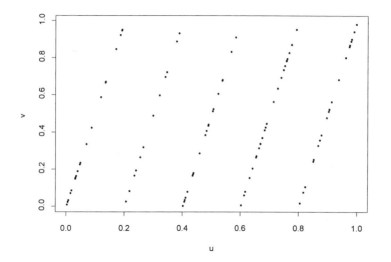

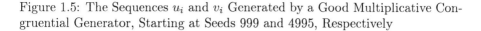

Figure 1.5: The Sequences u_i and v_i Generated by a Good Multiplicative Congruential Generator, Starting at Seeds 999 and 4995, Respectively

This would indicate that when choosing to skip ahead k steps, the seed should not be a small multiple of, or a small fraction of, k.

Another pair of sequences that may be of interest are ones that go in reverse order; that is, if one sequence is

$$x_1, \ x_2, \ x_3, \ \ldots, \ x_k, \ \ldots,$$

the other sequence is

$$x_k, \ x_{k-1}, \ x_{k-2}, \ \ldots, \ x_3, \ x_2, \ x_1, \ \ldots.$$

If the first sequence is generated by $x_{i+1} \equiv ax_i \bmod m$, the second sequence is generated by $y_{i+1} \equiv by_i \bmod m$, where

$$b \equiv a^{c-1} \bmod m,$$

and c is the period.

1.1.3   Shuffling the Output Stream

MacLaren and Marsaglia (1965) suggest that the output stream of a linear congruential random number generator be *shuffled* by using another, perhaps simpler, generator to permute subsequences from the original generator. This shuffling can increase the period (because it is no longer necessary for the same value to follow a given value every time it occurs) and it can also break up the lattice structure. (There will still be a lattice, of course; it will just have a different number of planes.)

Because a single random number can be used to generate independent random numbers ("bit stripping", see page 5), a single generator can be used to shuffle itself.

Bays and Durham (1976) suggest using a single generator to fill a table of length k and then using a single stream to select a number from the table and to replenish the table. After initializing a table T to contain $x_1, x_2, \ldots, x_k$, set $i = k + 1$ and generate x_i to use as an index to the table. Then update the table with x_{i+1}. The method is shown in Algorithm 1.1 to generate the stream y_i for $i = 1, 2, \ldots$.

Algorithm 1.1 Bays-Durham Shuffling of Uniform Deviates

   0. Initialize the table T with $x_1, x_2, \ldots, x_k$, $i = 1$, generate x_{k+i}, and set $y_i = x_{k+i}$.

   1. Generate j from y_i (use bit stripping or mod k).

   2. Set $i = i + 1$.

   3. Set $y_i = T(j)$.

   4. Generate x_{k+i}, and refresh $T(j)$ with x_{k+i}.   ∎

The period of the generator may be increased by this shuffling. Bays and Durham (1976) show that the period under this shuffling is $O(k!c)^{\frac{1}{2}}$, where c is the cycle length of the original, unshuffled generator. If k is chosen so that $k! > c$, the period is increased.

For example, with the generator used in Figure 1.2 ($m = 31$, $a = 3$ and beginning with $x_0 = 9$), which yielded the sequence

27, 19, 26, 16, 17, 20, 29, 25, 13, 8, 24, 10, 30, 28, 22, 4, 12, 5, 15, 14, 11, 2, 6, 18, 23, 7, 21, 1, 3, 9,

we select $k = 8$, and initialize the table as

$$27, 19, 26, 16, 17, 20, 29, 25.$$

We then use the next number, 13, as the first value in the output stream, and also to form a random index into the table. If we form the index as $13 \bmod 8 + 1$, we get the sixth tabular value, 20, as the second number in the output stream. We generate the next number in the original stream, 8, and put it in the table, so we now have the table

$$27, 19, 26, 16, 17, 8, 29, 25.$$

Now we use 20 as the index to the table and get the fifth tabular value, 17, as the third number in the output stream. By continuing in this manner to yield 10,000 deviates, and plotting the successive pairs, we get Figure 1.6. The very bad lattice structure shown in Figure 1.2 has diminished. (Remember there are only 30 different values, however.)

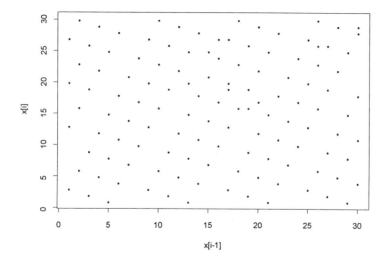

Figure 1.6: Pairs of Successive Numbers from a Shuffled Version of $x_i \equiv 3x_{i-1} \bmod 31$. Compare this with Figure 1.2

1.1.4 Tests of Linear Congruential Generators

Any number of statistical tests could be developed to be applied to the output of a given generator. The simple underlying idea is to form any transformation on the subsequence, determine the distribution of the transformation under the null hypothesis of independent uniformity of the sequence, and perform a goodness-of-fit test of that distribution. Useful transformations can be multivariate; the simplest is just the identity transformation on a subsequence of a chosen length. Another simple transformation is just to add successive terms.

(Adding two successive terms should yield a triangular distribution.) With a little imagination, the reader should be able to devise chi-squared tests for uniformity in any dimension, chi-squared tests for triangularity of sums of two successive numbers, and so on. Tests for serial correlation of various lags, and various sign tests are other possibilities that should come to mind to anyone with some training in statistics.

For some specific generators or families of generators there are extensive empirical studies reported in the literature. For $m = 2^{31} - 1$, for example, empirical studies by Fishman and Moore (1982, 1986) indicate that different values of multipliers, all of which perform well under the lattice test and the spectral test (see Section 6.1, page 151), may yield samples statistically distinguishable from samples from a true uniform distribution.

Park and Miller (1988) summarize some problems with random number generators commonly available, and propose a "minimal standard" for a linear congruential generator. The generator must perform "at least as well as" one with $m = 2^{31} - 1$ and $a = 16807$, which is a primitive root. (The smallest primitive root of $2^{31} - 1$ is 7; and $7^5 = 16807$ is the largest power of 7 such that $7^p x$, for the largest value of x (which is $2^{32} - 2$), can be represented in the common 64-bit floating point format.)

This choice of m and a is due to Lewis, Goodman, and Miller (1969), and is very widely used. It is implemented in the package by Learmonth and Lewis (1973a), and results of extensive tests by Learmonth and Lewis (1973b) are available for it. It is provided as one option in the IMSL Libraries. The value of 16807 for the multiplier was found by Fishman and Moore (1986) to be marginally acceptable, but there were several other multipliers that performed better in their battery of tests.

The article by Park and Miller has generated extensive discussion; see the "Technical Correspondence" in the July 1993, issue of *Communications of the ACM*, pages 105 through 110. It is relatively easy to program the minimal standard, as we see in the next section. The method described by Carta (1990), however, should be avoided because it does not implement the minimal standard generator.

Ferrenberg, Landau, and Wong (1992) used some of the generators that meet the Park and Miller minimal standard to perform several simulation studies in which the correct answer was known. Their simulation results suggested that even some of the "good" generators could not be relied on in some simulations. Vattulainen, Ala-Nissila, and Kankaala (1994) likewise used some of these generators as well as generators of other types and found that their simulations often did not correspond to the processes they were modeling. The point is that the "minimal standard" is *minimal*.

We discuss general empirical tests of random number generators in Section 6.2, page 154.

1.2 Computer Implementation of Linear Congruential Generators

There are many important issues to consider in writing computer software to generate random numbers. Before proceeding to discuss other types of random number generators, we consider the computer implementation of linear congruential generators. These same kinds of considerations apply to other generators as well.

As a problem in numerical analysis, the basic recurrence of the linear congruential generator is somewhat ill-conditioned. This is because full precision must be maintained throughout; if any term in the sequence is not exact in the last place, all subsequent terms will differ radically from the true sequence.

1.2.1 Insuring Exact Computations

Because of the number of significant digits in the quantities in this recurrence, even the novice programmer learns very quickly that special steps may be required to provide the complete precision required.

If the multiplier a is relatively large, a way of avoiding the need for higher precision is to perform the computations for $x_i \equiv ax_{i-1} \bmod m$ as

$$x_i \equiv (b(cx_{i-1} \bmod m) + dx_{i-1} \bmod m) \bmod m, \qquad (1.6)$$

where $a = bc + d$. The values of b, c, and d are chosen so that all products are exactly representable. This is possible as long as the available precision is at least $1.5\log_B m$ places in the arithmetic base B.

Even if a is relatively small, as in the Park and Miller (1988) "minimal standard", the computations cannot be performed directly in ordinary floating-point words on computers with 32-bit words. In C we can use double (fmod), and in Fortran we can use double precision, as shown in Figure 1.7.

```
      subroutine rnlcm (dseed, nr, u)
      double precision dseed, dm, dmp
      data dm/2147483647.d0/, dmp/2147483655.d0/
                 !(dmp is computer specific)
      do 10 i = 1, nr
         dseed = dmod (16807.d0*dseed, dm)
         u(i) = dseed/dmp
10    continue
      end
```

Figure 1.7: A Fortran Program Illustrating a Congruential Generator for a Machine with 32-Bit Words

Another way of avoiding the need for higher precision for relatively small multipliers, that is, multipliers smaller than $\sqrt{m}$ (as in the case shown in Figure 1.7), is to perform the computations in the following steps given by Bratley, Fox, and Schrage (1987). Let

$$q = \lfloor m/a \rfloor$$

and

$$r \equiv m \bmod a.$$

Then

$$
\begin{aligned}
ax_{i-1} \bmod m &= (ax_{i-1} - \lfloor x_{i-1}/q \rfloor m) \bmod m \\
&= (ax_{i-1} - \lfloor x_{i-1}/q \rfloor (aq + r)) \bmod m \\
&= (a(x_{i-1} - \lfloor x_{i-1}/q \rfloor)q - \lfloor x_{i-1}/q \rfloor r) \bmod m \\
&= (a(x_{i-1} \bmod q) - \lfloor x_{i-1}/q \rfloor r) \bmod m.
\end{aligned}
\tag{1.7}
$$

For the operations of equation (1.7), L'Ecuyer (1988) gives coding of the following form:

```
k = x/q
x = a*(x - k*q) - k*r
if (x < 0) x = x + m
```

As above, it is often convenient to use the fact that the modular reduction is equivalent to

$$x_i = ax_{i-1} - m \lfloor ax_{i-1}/m \rfloor.$$

But in this equivalent formulation, the floor operation must be performed simultaneously with the division operation (that is, there can be no rounding between the operations). This may not be the case in some computing systems, and this is just another numerical "gotcha" for this problem.

1.2.2   Restriction that the Output Be > 0 and < 1

There is another point that is often overlooked in general discussions of random number generation. We casually mentioned earlier in the chapter that we want to simulate a uniform distribution over $(0, 1)$. Mathematically, that is the same as a uniform distribution over $[0, 1]$; it is not the same on the computer, however. Using the methods we have discussed, we must sometimes be careful to exclude the endpoints. Whenever the mixed congruential generator is used (i.e., $c \neq 0$ in the recurrence (1.1)), we may have to take special steps to handle the case when $x_i = 0$. For the multiplicative congruential generator (1.2), we do not have to worry about a 0. The normalization $u_i = x_i/m$ will not yield a 0.

The normalization, however, can yield a 1, whether the generator is mixed or multiplicative. To avoid that we choose a different normalizer, $\tilde{m}$ ($> m$). See the code in Figure 1.7, and see Exercise 1.10, page 40.

1.2.3 Efficiency Considerations

In some computer architectures, operations on fixed-point numbers are faster than those on floating-point numbers. If the modulus is a power of 2, it may be possible to perform the modular reduction by simply retaining only the low-order bits of the product. Furthermore, if the power of 2 corresponds to the number of numeric bits, it may be possible to use the fixed-point overflow to do the modular reduction. This is an old "trick", which was implemented in many of the early generators, including RANDU. This trick can also be implemented using a bitwise "and" operation. (The Fortran intrinsic iand does this.) This can have the same effect but without causing the overflow. Overflow may be considered an arithmetic error. A power of 2 is generally not a good modulus, however, as we have already pointed out.

The fixed-point overflow trick can be modified for doing a modular reduction for $m = 2^p - 1$. Let

$$\tilde{x}_i \equiv (ax_{i-1} + c) \bmod (m+1),$$

and if $\tilde{x}_i < m + 1$, then

$$x_i = \tilde{x}_i;$$

otherwise,

$$x_i = \tilde{x}_i - m.$$

1.2.4 Vector Processors

Because a random number generator may be invoked millions of times in a single program, it is important to perform the operations efficiently. Brophy et al. (1989) describe an implementation of the linear congruential generator for a vector processor. If the multiplier is a and the modulus is m, and the vector register length is k, the quantities

$$a_j \equiv a^j \bmod m, \text{ for } j = 1, 2, \ldots, k,$$

are precomputed and stored in a vector register. The modulo reduction operation is not (usually) a vector operation, however. For the particular generator they considered, $m = 2^{31} - 1$, and because

$$\frac{x}{2^{31} - 1} = x 2^{-31} + x 2^{-62} + \cdots,$$

they did the modular reduction as $a_j x - (2^{31} - 1) \lfloor a_j 2^{-31} x \rfloor$. For a given x_i, the k subsequent deviates are obtained with a few operations in a vector register.

Anderson (1990) provides a survey of the methods for vector processors and other high-performance architectures.

How to implement a random number generator in a given environment depends on such things as

- the type of architecture (vector, parallel, etc.)

- the precision available

- whether integer arithmetic is available

- whether integer overflow is equivalent to modular reduction

- the base of the arithmetic

- the relative speeds of multiplication and division

- the relative speeds of MOD and INT.

The quality of the generator should never be sacrificed in the interest of efficiency, no matter what type of computer architecture is being used.

1.3 Other Congruential Generators

There are several other variations of the basic congruential generator. They share the fundamental operation of modular reduction that provides the "randomness" or the "unpredictability". In general, the variations that have been proposed increase the complexity of the generator. For a generator to be useful, however, we must be able to analyze and understand its properties at least to the point of having some assurance that there is no deleterious nonrandomness lurking in the output of the generator. We should be also able to skip ahead in the generator a fixed distance. More study is needed for most of the generators mentioned in this section.

1.3.1 Multiple Recursive Generators

A simple extension of the multiplicative congruential generator is to use multiples of the previous k values to generate the next one:

$$x_i \equiv (a_1 x_{i-1} + a_2 x_{i-2} + \cdots a_k x_{i-k}) \bmod m.$$

This is sometimes called a "multiple recursive" multiplicative congruential generator. A multiple recursive generator can have a much longer period than a simple multiplicative generator. L'Ecuyer, Blouin, and Couture (1993) study some of the statistical properties of multiple recursive generators, and make recommendations for the multipliers for some specific moduli, including the common one, $2^{31} - 1$. Their assessment of the quality of the generators is based on the Beyer ratio (see page 153). They also discuss programming issues for such generators, and give code for a short C program for a multiple recursive generator with $k = 5$. Their code is portable and will not return 0 or 1. They also indicate how to skip ahead in multiple recursive generators.

1.3.2 Lagged Fibonacci

A simple Fibonacci sequence has $x_{i+2} = x_{i+1} + x_i$. Reducing the numbers mod m produces a sequence that looks somewhat random, but actually does not have satisfactory randomness properties.

We can generalize the Fibonacci recursion in two ways. First, instead of combining successive terms, we combine terms at some greater distance apart, so the *lagged Fibonacci congruential generator* is

$$x_i \equiv (x_{i-j} + x_{i-k}) \bmod m. \tag{1.8}$$

A lagged Fibonacci congruential generator requires an initial sequence, rather than just a single seed.

If j, k, and m are chosen properly, the lagged Fibonacci generator can perform well. If m is a prime and $k > j$, the period can be as large as $m^k - 1$. More important, this generator is the basis for other generators, such as discussed in Section 1.5, page 30.

Altman (1989) has shown that care must be exercised in selecting the initial sequence, but for a carefully chosen initial sequence, the bits in the output sequence seem to have better randomness properties than those from congruential and Tausworthe generators (page 26).

A second way of generalizing the Fibonacci recursion is to use more general binary operators, instead of addition modulo m. In a *general* lagged Fibonacci generator we start with $x_1, x_2, \ldots, x_k$ ("random" numbers), and let

$$x_i = (x_{i-j} \circ x_{i-k}),$$

where $\circ$ is some binary operator, with

$$0 \le x_i \le m - 1, \quad \text{and} \quad 0 < j < k < i.$$

1.3.3 Add-with-Carry, Subtract-with-Borrow, and Multiply-with-Carry Generators

Marsaglia and Zaman (1991) describe two variants of a generator they called "add-with-carry" (AWC) and "subtract-with-borrow" (SWB). The add-with-carry form is

$$x_i \equiv (x_{i-s} + x_{i-r} + c_i) \bmod m,$$

where $c_1 = 0$, and $c_{i+1} = 0$ if $x_{i-s} + x_{i-r} + c_i < m$, and $c_{i+1} = 1$ otherwise. The c is the "carry".

Marsaglia and Zaman investigated various values of s, r, and m. For some choices they recommended, the period of the generator is of the order of 10^{43}. This generator can also be implemented in modulo 1 arithmetic. Tezuka and L'Ecuyer (1992) have shown that sequences resulting from this generator are essentially equivalent to sequences from linear congruential generators with very large prime moduli. (In fact, the AWC/SWB generators can be viewed as efficient ways of implementing such large linear congruential generators.) The work of Couture and L'Ecuyer (1994) and Tezuka, L'Ecuyer, and Couture (1994) indicates that the lattice structure in high dimensions may be very poor.

Marsaglia also described a multiply-with-carry random number generator that is a generalization of the add-with-carry random number generator. The multiply-with-carry generator is

$$x_i \equiv (ax_{i-1} + c_i) \bmod m,$$

Marsaglia suggests $m = 2^{32}$ and an implementation in 64-bit integers. The use of the lower-order 32 bits results in the modular reduction, as we have seen. The higher-order 32 bits determine the carry. Couture and L'Ecuyer (1995) suggest a method for finding parameters for the multiply-with-carry generator that yield good sequences.

1.3.4 Inversive Congruential Generators

Inversive congruential generators, introduced by Eichenauer and Lehn (1986), use the modular multiplicative inverse (if it exists) to generate the next variate in a sequence. The inversive congruential generator is

$$x_i \equiv (ax_{i-1}^- + c) \bmod m, \quad \text{with} \quad 0 \le x_i < m, \tag{1.9}$$

where x^- denotes the multiplicative inverse of x modulo m, if it exists, or else it denotes 0. The multiplicative inverse, x^-, of x modulo m is defined for all $x \neq 0$ by

$$1 \equiv x^- x \bmod m.$$

Compare this generator with (1.1), page 6. Eichenauer and Lehn (1986); Eichenauer, Grothe, and Lehn (1988); and Niederreiter (1988, 1989) show that the inversive congruential generators have good uniformity properties, in particular with regard to lattice structure and serial correlations.

Eichenauer-Herrmann and Ickstadt (1994) introduced the *explicit inversive congruential* method that yields odd integers between 0 and $m - 1$; and Eichenauer-Herrmann (1996) gives a modification of it that can yield all nonnegative integers up to $m - 1$. The modified explicit inversive congruential generator is given by

$$x_i \equiv i(ai + b)^- \bmod 2^p, \quad i = 0, 1, 2, \ldots$$

where $a \equiv 2 \bmod 4$, $b \equiv 1 \bmod 2$, and $p \ge 4$. Eichenauer-Herrmann (1996) shows that the period is m and the 2-d discrepancy (see Section 6.1, page 151) is $O(m^{-\frac{1}{2}}(\log m)^2)$ for the modified explicit inversive generator.

Although there are some ways of speeding up the computations (see Gordon, 1989), the computational difficulties of the inversive congruential generators have prevented their widespread use, but they hold some promise as high-quality generators. Chou and Niederreiter (1995) describe a lattice test and some results of it for inversive congruential generators. An inversive congruential generator does not yield regular planes as a linear congruential generator. Leeb and Wegenkittl (1997) report on several tests of inversive congruential generators. The apparent randomness was better than that of the linear congruential generators they tested.

1.3.5 Other Nonlinear Congruential Generators

A simple generalization of the linear congruential generator (1.1) is suggested by Knuth (1981):

$$x_i \equiv (dx_{i-1}^2 + ax_{i-1} + c) \bmod m, \quad \text{with} \quad 0 \le x_i < m.$$

Higher-degree polynomials could be used also. Whether there is any advantage in using higher degree polynomials is not clear. They do have the obvious advantage of making the generator less obvious.

Blum, Blum, and Shub (1986) proposed a generator based on

$$x_i \equiv x_{i-1}^2 \bmod m.$$

Their method, however, goes beyond the introduction of just more complicated transformations in the sequence of integers. They form the output sequence as the bits $b_1 b_2 b_3 \cdots$, where $b_i = 0$, if x_i is even, and otherwise $b_i = 1$. Blum, Blum, and Shub showed that if $m = p_1 p_2$, where p_1 and p_2 are distinct primes, each congruent to 3 mod 4, then the output of this generator is not predictable in polynomial time without knowledge of p_1 and p_2. Because of this unpredictability, this generator has important possible uses in cryptography.

A more general congruential generator uses a function g of the previous value to generate the next one:

$$x_i \equiv g(x_{i-1}) \bmod m, \quad \text{with} \quad 0 \le x_i < m. \tag{1.10}$$

Such generators were studied by Eichenauer, Grothe, and Lehn (1988). The wide choice of functions can yield generators with very good uniformity properties, but there is none other than the simple linear or inversive ones for which a body of theory and experience would suggest its use in serious work.

Kato, Wu, and Yanagihara (1996a) consider a function g in (1.10) that combines a simple linear function with the multiplicative inverse in (1.9). Their generator is

$$x_i \equiv (ax_{i-1}^- + bx_{i-1} + c) \bmod m, \quad \text{with} \quad 0 \le x_i < m.$$

They suggest a modulus that is a power of two, and in Kato, Wu, and Yanagihara (1996b) derive properties for this generator similar to those derived by Niederreiter (1989) for the inversive generator in (1.9).

Eichenauer-Herrmann (1995) provides a survey of work done on nonlinear generators, including the inversive congruential generators.

1.3.6   Matrix Congruential Generators

A generalization of the scalar linear congruential generator to a generator of pseudorandom vectors is straightforward:

$$x_i \equiv (Ax_{i-1} + C) \bmod m,$$

where the x_i are vectors of length d, and A and C are $d \times d$ matrices. The elements of the vectors and of the matrices are integers between 1 and $m - 1$. The vector elements are then scaled into the interval $(0, 1)$ to simulate $U(0, 1)$ deviates. Such a generator is called a *matrix congruential generator*. As with the scalar generators, C is often chosen as 0 (a matrix with all elements equal to 0). Reasons for using a matrix generator are to generate parallel streams of pseudorandom deviates or to induce a correlational structure in the random vectors.

If A is a diagonal matrix, the matrix generator is essentially a set of scalar generators with different multipliers. For more general A, the elements of the vectors are correlated. Rather than concentrating directly on the correlations, most studies of the matrix generators (e.g., Afflerbach and Grothe, 1988) have focused on the lattice structure. Choosing A as the Cholesky factor of a target variance-covariance matrix may make some sense, and there may be other situations in which a matrix generator would be of some value. In Exercise 1.3, page 39, you are asked to experiment with a matrix generator to study the correlations. Generally, however, any desired correlational structure should be simulated by transformations on an independent stream of uniforms, as we discuss in Section 3.2, page 105, rather than trying to induce it in the congruential generator.

Analysis of the period of the matrix congruential generator is somewhat more complicated than that of the scalar generator (see Grothe, 1987). It is clear that A should be nonsingular because otherwise it is possible to generate a zero vector, at which point all subsequent vectors are zero.

A straightforward extension of the matrix congruential generator is the multiple recursive matrix random number generator:

$$x_i \equiv (A_1x_{i-1} + A_2x_{i-2} + \cdots A_kx_{i-k}) \bmod m.$$

This is the same idea as in the multiple recursive generator for scalars, as considered above. Niederreiter (1993, 1995a, 1995b, 1995d) discusses some of its properties.

1.4   Feedback Shift Register Generators

Tausworthe (1965) introduced a generator based on a sequence of 0's and 1's generated by a recurrence of the form

$$a_i \equiv (c_pa_{i-p} + c_{p-1}a_{i-p+1} + \cdots + c_1a_{i-1}) \bmod 2, \qquad (1.11)$$

where all variables take on values of either 0 or 1. A similar generator could, of course, be devised with a modulus m, not necessarily 2.

If the modulus is a prime, the generator can be related to a polynomial

$$f(z) = z^p - (c_1 z^{p-1} + \cdots + c_{p-1} z + c_p) \tag{1.12}$$

over the Galois field defined over the integers $0, 1, \ldots, m-1$ with the addition and multiplication being defined in the usual way followed by a reduction modulo m. We will denote a Galois field over a set with m elements as $\mathbb{G}(m)$. An important result from the theory developed for such polynomials is that, so long as the initial vector of a's are not all 0's, the period of the recurrence (1.11) is $m^p - 1$ if and only if the polynomial (1.12) is irreducible (i.e., cannot be factored) over $\mathbb{G}(m)$. (An irreducible polynomial is also called a primitive polynomial.)

It is easy to see that the maximal period is $m^p - 1$, because if any p-vector of a's repeats, all subsequent values are repeats. If $p = 1$, we have the multiplicative congruential generator. See Knuth (1981) for further discussion of these properties as well as for a method to determine primitive polynomials.

For computational efficiency, the modulus in (1.11) should be 2 and most of the c's should be zero. For $m = 2$, there is only one binomial that is irreducible; and it is $z + 1$, which would yield an unacceptable period. There are, however, many trinomials (see Zierler and Brillhart, 1968, for a list of all primitive trinomials modulo 2 up to degree 1,000). Hence, the recurrence (1.11) often has the form

$$a_i \equiv (a_{i-p} + a_{i-p+q}) \bmod 2, \tag{1.13}$$

resulting from a trinomial. Addition of 0's and 1's modulo 2 is the binary exclusive-or operation, denoted by $\oplus$; thus we may write the recurrence as

$$a_i = a_{i-p} \oplus a_{i-p+q}. \tag{1.14}$$

After this recurrence has been evaluated a sufficient number of times, say l (with $l \leq p$), the l-tuple of a's is interpreted as a number in base 2. This is referred to as an l-wise decimation of the sequence of a's. If l is relatively prime to $2^p - 1$ (the period of the sequence of a's), the period of the l-tuples will also be $2^p - 1$. In this case the decimation is said to be proper. Note that the recurrence of bits is the same recurrence of l-tuples,

$$x_i = x_{i-p} \oplus x_{i-p+q}, \tag{1.15}$$

where the x's are the numbers represented by interpreting the l-tuples as binary notation, and the exclusive-or operation is performed bit-wise.

As an example, consider the trinomial

$$x^4 + x + 1, \tag{1.16}$$

and begin with the bit sequence

$$1, 0, 1, 0.$$

For this polynomial, $p = 4$ and $q = 3$ in the recurrence (1.14). Operating the generator, we obtain

$$1, 1, 0, 0, 1, 0, 0, 0, 1, 1, 1, 1, 0, 1, 0,$$

at which point the sequence repeats; its period is $2^4 - 1$. A 4-wise decimation using the recurrence (1.15) yields the numbers

$$12, 8, 15, 5, \ldots$$

(in which the 5 required an additional bit in the sequence above). We could continue in this way to get 15 (that is, $2^4 - 1$) integers between 1 and 15 before the sequence began to repeat.

As with the linear congruential generators, different values of the c's and even of the starting values of the a's can yield either good generators (i.e., ones whose outputs seem to be random samples of a uniform distribution) or bad generators.

This recurrence operation can be performed in a *feedback shift register*, which is a vector of bits that is shifted, say, to the left, one bit at a time, and the bit shifted out is combined with other bits in the register to form the rightmost bit. The operation can be pictured as shown in Figure 1.8, where the bits are combined using $\oplus$.

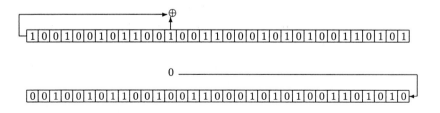

Figure 1.8: One Shift of a Feedback Shift Register

1.4.1   Generalized Feedback Shift Registers and Variations

Feedback shift registers have an extensive theory (see, e.g., Golomb, 1982) because they have been used for some time in communications and cryptography.

A variation on the Tausworthe generator is suggested by Lewis and Payne (1973), who call their modified generator a *generalized feedback shift register* (GFSR) generator. In the GFSR method, the bits of the sequence from recurrence (1.13) form the bits in one binary position of the numbers being generated. The next binary position of the numbers being generated is filled with the same bit sequence but with a delay. By using the bit stream from the trinomial $x^4 + x + 1$ and the starting sequence we considered before, and again forming

4-bit words by putting the bits into a fixed binary position with a delay of 3 between binary positions, we have

$$
\begin{aligned}
1010 &= 10 \\
1110 &= 14 \\
0011 &= 3 \\
0101 &= 5 \\
1111 &= 15 \\
0001 &= 1 \\
0010 &= 2 \\
0111 &= 7 \\
1000 &= 8 \\
1001 &= 9 \\
1011 &= 11 \\
1100 &= 12 \\
0100 &= 4 \\
1101 &= 13 \\
0110 &= 6
\end{aligned}
$$

at which point the sequence repeats. Notice that the recurrence (1.15),

$$ x_i = x_{i-p} \oplus x_{i-p+q}, $$

still holds, where the x's are the numbers represented by interpreting the l-tuples as binary notation, and the exclusive-or operation is performed bit-wise.

Lewis and Payne (1973) discussed methods of initializing the generator, and gave programs both for the initialization and for the generation. Kirkpatrick and Stoll (1981) presented a faster way of initializing the generator and developed a program, R250, implementing a generator with $p = 250$. This program is widely used among physicists.

Fushimi (1990) has studied GFSR methods in general, and has described some particular GFSR's, which he showed to have good properties, both theoretically and empirically. One particular case studied by Fushimi had $p = 521$, with $c_{521} = c_{32} = 1$ (with a slight modification in the definition of the recursion as $x_i = x_{i-3p} \oplus x_{i-3q}$). The generator, which is available in the IMSL Libraries, has a period of $2^{521} - 1$.

Matsumoto and Kurita (1992, 1994) propose a modification of the GFSR in the recurrence (1.15), by "twisting" the bit pattern in x_{i-p+q}. This is done by viewing the x's as l-vectors of zeros and ones and multiplying x_{i-p+q} by an $l \times l$ matrix A. The recurrence then becomes

$$ x_i = x_{i-p} \oplus A x_{i-p+q}. $$

They call this a *twisted GSFR generator*. Matsumoto and Kurita (1994) show how to choose A and modify the output so as to achieve good uniformity properties. They also gave a relatively short C program implementing the method. Further empirical studies are needed for this generator. The idea of a twisted

GFSR generator is related to the matrix congruential generators discussed in
Section 1.3.6.

In our discussion of computer implementation of congruential generators
beginning on page 19, we mentioned a technique to insure that the result of
the generator be less than 1. The same technique, of course, works for GFSR
generators, but in these generators we have the additional concern of a 0. By
using the obvious bit operations, which are inexpensive, it is usually fairly easy
to prevent 0 from occurring, however.

1.4.2 Skipping Ahead in GFSR Generators

Golomb (1982) noted that the basic recurrence (1.15),

$$x_i = x_{i-p} \oplus x_{i-p+q},$$

implies

$$x_i = x_{i-2^e p} \oplus x_{i-2^e p + 2^e q},$$

for any integer e. This provides a method of skipping ahead in a generalized
feedback shift register generator by a fixed distance that is a power of 2. Aluru,
Prabhu, and Gustafson (1992) described a leapfrog method using this relationship and applied it to parallel random number generators. So for k a power of
2, we have

$$x_i = x_{i-kp} \oplus x_{i-k(p-q)},$$

and so we can generate a leapfrog sequence analogous to (1.5), page 14, for
multiplicative congruential generators:

$$x_s, x_{s+k}, x_{s+2k}, \cdots$$

1.5 Other Sources of Uniform Random Numbers

Many different mechanisms for generating random numbers have been introduced. Some are simple variations of the linear congruential generator or the
feedback shift register, perhaps designed for microcomputers, or some other
special environment. Various companies distribute proprietary random number generators, perhaps based on some physical process. Marsaglia (1995) uses
some of those, and his assessment of the ones he used was that they were not of
high quality. (Nevertheless, he combined them with other sources and produced
pseudorandom streams of high quality.)

There are some connections between random number generators and chaotic
systems. Chaotic systems generally cannot be used to generate useful random
processes, but some of the results of chaos theory may have some relevance to
the understanding of pseudorandom generators.

1.5.1 Generators Based on Chaotic Systems

Lüscher (1994) relates the Marsaglia-Zaman subtract-with-borrow generator to a chaotic dynamical system. This relationship allows an interpretation of the correlational defects of the basic generator as short-term effects, and provides a way for skipping ahead in the generator to avoid those problems. The Lüscher generator has a long period and performed well in the statistical tests reported by Lüscher. James (1994) provides a portable Fortran program, RANLUX, implementing this generator.

1.5.2 Tables of Random Numbers

Before the development of reliable methods for generating pseudorandom numbers, there were available a number of tables of random numbers that had been generated by some physical process or that had been constructed by systematically processing statistical tables to access midorder digits of the data they contained. The most widely used of these were the "RAND tables" (RAND Corporation, 1955), which contain a million random digits generated by a physical process. The random digits in these tables have been subjected to many statistical tests, with no evidence that they did not arise from a discrete uniform distribution over the ten mass points.

Marsaglia (1995) has produced a CD-ROM that contains 4.8 billion random bits stored in sixty separate files. The random bits were produced by a combination of two or more deterministic random number (pseudorandom number) generators and three devices that use physical processes. Some of the files also have added bit streams from digitized music or pictures.

1.6 Portable Random Number Generators

A major problem with random number generation on computers is that programs for the same generators on different computers often yield different sequences. For programs that make use of random number generators to be portable, the generator must produce the same sequence in all computer/compiler environments. It is desirable that the generators be portable to facilitate transfer of research and development efforts. Portability reduces the number of times the wheel is reinvented as well as the amount of the computer-knowledge overhead that burdens a researcher. The user can devote attention to the research problem rather than to the extraneous details of the computer tools used to address the problem.

The heterogeneous computing environment in which most statisticians work has brought an increased importance to portability of software. Formerly, portability was a concern primarily for distributors of software, for users who may be switching jobs, or for computer installations changing or contemplating changing their hardware. With the widespread availability of personal computers, all computer users now are much more likely to use (or to attempt to use) the same program on more than one machine. There are both technical and tactical reasons for using a micro and a mainframe while working on the same problem. The technical reasons include the differences in resources (memory, CPU speed, software) available on micros and mainframes. These differences likely will continue. As new and better micros are introduced and more software is developed for them, new and better supercomputers will also be developed.

The availability of the different computers in different working environments such as home, lab, and office means that using multiple computers on a single problem can make more efficient use of one's time. These tactical reasons for using multiple computers will persist, and it will become increasingly commonplace for a researcher to use more than one computer.

Many programming languages and systems come with built-in random number generators. The quality of the built-in generators varies widely (see Lewis and Orav, 1989, for analyses of some of these generators). Generators in the same software system, such as `rand()` in `stdlib.h` of the C programming language, may not generate the same sequence on different machines or even in different C compilers on the same machine. It is generally better to use a random number generator from a system such as the IMSL Library, which provides portability across different platforms.

The algorithms for random number generation are not always straightforward to implement, as we have discussed on page 19. Algorithms with relatively small operands, such as in Wichmann and Hill (1982) and the alternate generator of L'Ecuyer (1988), are likely to be portable; and, in fact, can even be implemented on computers with 16-bit words. The multipliers suggested by Wu (1997) that we discussed above can fairly easily be implemented portably.

We have mentioned several subtle problems for implementing congruential generators on page 19. Other generators have similar problems, such as the 0 and 1 problem, of which most people who have never built generators used in large-scale simulations are not aware. See Gentle (1981) for further discussion of the issue of portability and how it can be achieved in random number generators.

1.7   Combining Generators

Both the period and the apparent randomness of random number generators can often be improved by combining more than one generator. The shuffling methods of MacLaren and Marsaglia (1965) and others, described on page 16, may use two or more generators in filling a table and shuffling it.

Another way of combining generators is suggested by Collings (1987). His method requires a pool of generators, each maintaining its own stream. For each number in the delivered stream, an additional generator is used to select which of the pool of generators is to be used to produce the number.

The generators that are combined can be of any type. One of the early combined generators, developed by George Marsaglia and called "Super-Duper", used a linear congurential generator and a Tauseworthe generator. See Learmonth and Lewis (1973b) for a description of Super-Duper.

1.7.1 Wichmann/Hill Generator

Wichmann and Hill (1982 and corrigendum, 1984) describe a combination of three linear congruential generators that is easy to program and has good randomness properties. The generator is

$$x_i \equiv 171x_{i-1} \bmod 30269$$
$$y_i \equiv 172y_{i-1} \bmod 30307$$
$$z_i \equiv 170z_{i-1} \bmod 30323,$$

and

$$u_i = \left(\frac{x_i}{30269} + \frac{y_i}{30307} + \frac{z_i}{30323} \right) \bmod 1.$$

This generator requires three seeds (x_0, y_0, and z_0) and yields numbers u_i in the interval (0,1). The period is of the order of 10^{12}. Notice that the final modular reduction is equivalent to retaining only the fractional part of the sum of three uniform numbers. (See Exercise 1.1a, page 38.)

De Matteis and Pagnutti (1993) studied the higher-order autocorrelations in sequences from the Wichmann/Hill generator and found them to compare favorably with those from other good generators. This, together with the ease of implementation, makes the Wichmann/Hill generator a useful one for common applications. A straightforward implementation of this generator can yield a 0. (See Exercise 1.13, page 40.)

1.7.2 L'Ecuyer Combined Generators

L'Ecuyer (1988) suggests combining k multiplicative congruential generators that have prime moduli m_j, such that $(m_j - 1)/2$ are relatively prime, and with multipliers that yield full periods. Let the sequence from the jth generator be $x_{j,1}, x_{j,2}, x_{j,3}, \ldots$. Assuming that the first generator is a relatively "good" one, and that m_1 is fairly large, we form the ith integer in the sequence as

$$x_i = \sum_{j=1}^{k} (-1)^{j-1} x_{j,i} \bmod (m_1 - 1).$$

The other moduli m_j do not need to be large.

The normalization takes care of the possibility of a 0 occurring in this sequence:

$$u_i = \begin{cases} x_i/m_1 & \text{if } x_i > 0, \\ (m_1 - 1)/m_1 & \text{if } x_i = 0. \end{cases}$$

A specific generator suggested by L'Ecuyer (1988) is the following:

$$
\begin{aligned}
x_i &\equiv 40014 x_{i-1} \bmod 2147483563 \\
y_i &\equiv 40692 y_{i-1} \bmod 2147483399 \\
z_i &= x_i - y_i
\end{aligned}
$$

where if $z_i < 1$ then $z_i = z_i + 2147483562$, and finally,

$$
u_i = 4.656613 z_i \times 10^{-10}.
$$

The period is of the order of 10^{18}. L'Ecuyer presented results of both theoretical and empirical tests that indicate that the generator performs well.

Notice that the difference of two discrete uniforms is also a discrete uniform. (See Exercise 1.1c, page 38.) Notice also the normalization always yields numbers less than 1 because the normalizing constant is larger than 2147483563.

L'Ecuyer (1988) gives a portable program for the generator that uses the techniques we have discussed for keeping the results of intermediate computations small (see page 19 and following).

1.7.3   Properties of Combined Generators

We have mentioned the important difference between the ranges $[0, 1]$ and $(0, 1)$ for the simulated uniform distribution. Combining generators exposes the naive implementer to this problem anew. Whenever two generators are combined, there is a chance of obtaining a 0 or a 1 even if the output of each is over $(0, 1)$. See Exercise 1.13.

If the streams in a combination generator suffer similar irregularities in their patterns, the combination may not be able to overcome the problems. Combining some generators can actually degrade the quality of the output. Generally a judicious combination of two generators will improve the randomness properties, however. Consider two sequences produced by random number generators over a finite set S, which without loss we can consider to be the integers $1, 2, 3, \ldots, n$. A result due to Marsaglia (1985) suggests that the distribution of a sequence produced by a one-to-one function from $S \times S$ onto S is at least as close to uniform as the distribution of either of the original sequences.

Let $p_1, p_2, p_3, \ldots, p_n$ be the probabilities of obtaining $1, 2, 3, \ldots, n$ associated with the first sequence corresponding to the random variable X, and $q_1, q_2, q_3, \ldots, q_n$ be the corresponding probabilities associated with the second sequence that are realizations of the random variable Y. The distance of either sequence from the uniform distribution is taken to be the Euclidean norm of the vector of differences of the probabilities from the uniform $1/n$. Let s be the vector with all n elements equal to $1/n$, and let p be the vector with elements p_i. Then, for example,

$$
d(X) = \|p - s\| = \sqrt{\sum \left(p_i - \frac{1}{n} \right)^2}.
$$

This is similar to the discrepancy measure discussed in Section 6.1, page 151. It can be thought of as a measure of uniformity. The one-to-one function from $S \times S$ onto S can be represented as the operator $\circ$:

			X			
$\circ$	1	2	3	$\cdots$	n
1	$z_{(1,1)}$	$z_{(1,2)}$	$z_{(1,3)}$	$\cdots$	$z_{(1,n)}$
2	$z_{(2,1)}$	$z_{(2,2)}$	$z_{(2,3)}$	$\cdots$	$z_{(2,n)}$
Y 3	$z_{(3,1)}$	$z_{(3,2)}$	$z_{(3,3)}$	$\cdots$	$z_{(3,n)}$
$\cdots$	$\cdots$	$\cdots$	$\cdots$	$\cdots$	$\cdots$
n	$z_{(n,1)}$	$z_{(n,2)}$	$z_{(n,3)}$	$\cdots$	$z_{(n,n)}$

that is, $i \circ j = z_{(i,j)}$. In each row and in each column every integer from 1 to n occurs as the value of some $z_{(i,j)}$ exactly once. The probability vector for the random variable $Z = X \circ Y$ is then given by

$$\Pr(Z = k) = \sum_i \Pr(X = i)\Pr(Y = j \circ k),$$

where j is such that $i \circ j = k$. This expression can be written as Mq, where M is the matrix whose rows m_i^T correspond to permutations of the elements $p_1, p_2, \ldots, p_n$, such that each p_j occurs in only one column, and q is the vector containing the elements q_j. Now

$$\sum_i \Pr(Y = j \circ k) = 1,$$

so each row (and also each column) of M sums to 1, and $Ms = s$. For such a matrix M we have for any vector $v = (v_1, v_2, \ldots, v_n)$,

$$
\begin{aligned}
\|Mv\|^2 &= \left(\sum m_{1i}v_i\right)^2 + \left(\sum m_{2i}v_i\right)^2 + \cdots \left(\sum m_{ni}v_i\right)^2 \\
&\leq \sum m_{1i}v_i^2 + \sum m_{2i}v_i^2 + \cdots \sum m_{ni}v_i^2 \\
&= \sum v_i^2 \\
&= \|v\|^2
\end{aligned}
\tag{1.17}
$$

(see Exercise 1.14, page 40). Then we have

$$
\begin{aligned}
d(X \circ Y) &= \|Mq - s\| \\
&= \|M(q - s)\| \\
&\leq \|q - s\| \\
&= d(Y).
\end{aligned}
$$

Heuristically, this increased uniformity implies greater "uniformity" of the subsequences of $X \circ Y$ than of subsequences of either X or Y.

1.8 Independent Streams and Parallel Random Number Generation

Many of the Monte Carlo methods are "embarrassingly" parallel; they consist of independent computations, at least at some level, that are then averaged. The main issue for Monte Carlo methods performed in parallel is that the individual computations (that is, all computations except the outer averaging loop) be performed independently. For the random number generators providing data for parallel Monte Carlo computations, the only requirement over and above those of any good random number generator is to satisfy this requirement of independence of the random number streams. This is exactly the same problem as blocking or renewal, which has long been a common practice in simulation (see, e.g., Mihram, 1972, and Maclaren, 1989). Blocks are used in simulation experiments in much the same way as in ordinary experiments: to obtain a better measure of the underlying variation. In each block or each simulat`_`on run, an independent stream of random numbers is required.

Although several methods have been proposed for generating independent streams of random numbers, there are basically two kinds of methods. One type is based on skipping ahead in the stream, either by using different starting points or by a leapfrog method, as discussed in Section 1.1 on page 14 for linear congruential generators and in Section 1.4 on page 30 for GFSR generators. Both kinds of skipping ahead can be done randomly or at a fixed distance. The other type of method uses a combination of generators, either by simply using completely different generators in each of the parallel streams or by using a combination, such as suggested in Section 1.7, page 32.

1.8.1 Lehmer Trees

Frederickson et al. (1984) describe a way of combining linear congruential generators to form what they call a Lehmer tree, which is a binary tree in which all right branches or all left branches form a sequence from a Lehmer linear congruential generator. The tree is defined by the two recursions, both of which are the basic recursion (1.1) introduced by Lehmer,

$$x_i^{(L)} \equiv (a_L x_{i-1} + c_L) \bmod m,$$

and

$$x_i^{(R)} \equiv (a_R x_{i-1} + c_R) \bmod m.$$

At the $(i-1)^{\text{th}}$ node in the tree, an ordinary Lehmer sequence with seed x_{i-1} is generated by all right-branch nodes below it. A new sequence is initiated by taking a left branch to the first node below it and then all right branches from then on. The question is whether a (finite) sequence of all right-branch nodes below a given node is independent from the (finite) sequence of all right-branch nodes below the left-branch node immediately below the given node.

("Independent" for these finite subsequences can be interpreted strictly as having no elements in common.) Frederickson et al. gave conditions on a_L, c_L, a_R, c_R, and m that would guarantee the independence for a fixed length of the sequences.

Although Frederickson et al. (1984) concentrate on congruential generators whose modulus is a power of 2, Bowman and Robinson (1987) show that Lehmer trees constructed from generators using a modulus of that form had very poor properties. The problem is the same that we pointed out in Section 1.1, page 6, for generators with such moduli; the low-order bits have very short periods.

If the modulus in the lagged Fibonacci generator (1.8), page 23,

$$x_i \equiv (x_{i-j} + x_{i-k}) \bmod m,$$

is a power of 2, say 2^p, the maximum period possible is $(2^k - 1)2^{p-1}$. Mascagni et al. (1995) describe a method of identifying and using exclusive substreams, each with the maximal possible period.

Neither the congruential generator nor the generalized feedback register generator has entirely satisfactory methods of skipping ahead for use in parallel random number generation. As we have already mentioned, the congruential generator also suffers from a relatively short period.

1.8.2 Combination Generators

The ability to obtain independent streams and the long period and other good properties of a GFSR generator can be achieved by a combination generator in which a simple skip-ahead method is implemented in one of the generators, either a congruential or a GFSR generator, and the other generator is a good GFSR generator. Wollan (1992) describes such a method, which combines a skipping multiplicative congruential generator with a lagged Fibonacci generator, by subtraction modulo 1 of the normalized output. For the skipping congruential generator based on $x_i \equiv ax_{i-1} \bmod m$, the seed $a^j x_0$ is used for the jth processor, and the recurrence $x_i \equiv a^k x_{i-1} \bmod m$ is used in each processor. The same Fibonacci generator is used in all processors. Wollan also allowed for a single process to spawn a child process. Both the parent and the child process begin using a^2 in place of a (i.e., the multiplier for both is a^{2k} when one of the original k processes spawns a new process), and the child process skips from the current value x_i to ax_i.

1.8.3 Monte Carlo on Parallel Processors

Eddy (1990) and Mascagni et al. (1993) provide overviews of some of the issues and some of the proposed methods, especially those making use of linear congruential generators. Anderson (1990) discusses general methods for high-performance architectures. His emphasis is on vector processors, however.

Assessing the quality of random number generators operating in parallel can be quite difficult. Hanxleden and Scott (1992) discuss ways determining correctness of a random number generator operating in parallel, and they describe a dynamic way of choosing seeds in the different processors. ("Correctness" in this case means that the computer is executing a program that implements the algorithm correctly.) See also Cuccaro, Mascagni, and Pryor (1994) for some of the issues involved.

Exercises

1.1. Modular reduction and uniform distributions.

   (a) Let R be a random variable with a $U(0,1)$ distribution, let k be a nonzero integer constant, and let c be a real constant. Let

   $$S \equiv (kR + c) \bmod 1, \text{ with } 0 \leq S \leq 1.$$

   Show that S has a $U(0,1)$ distribution. *Hint:* First let $c = 0$ and consider kR; then consider $T + c$, where T is from $U(0,1)$.

   (b) Prove a generalization of Exercise 1.1a in which the constant c is replaced by a random variable with any distribution.

   (c) Let T be a random variable with a discrete uniform distribution with mass points $0, 1, \ldots, d-1$. Let $W_1, W_2, \ldots, W_n$ be independently distributed as discrete uniform random variables with integers as mass points. Show that

   $$T + \sum_{i=1}^{n} W_i \bmod d$$

   has the same distribution as T. (The reduced modulus is used in this expression, of course.) *Hint:* First consider $T + W_1$, and write $(T + W_1) \bmod d$ as $j + kd$, where j and k are integers with $0 \leq j \leq d-1$. (See also L'Ecuyer, 1988.)

1.2. Use Fortran, C, or some other programming system to write a program to implement a generator using a multiplicative congruential method with $m = 2^{13} - 1$ and $a = 17$. Generate 500 numbers x_i. Compute the correlation of the pairs of successive numbers, x_{i+1} and x_i. Plot the pairs. On how many lines do the points lie? Now let $a = 85$. Generate 500 numbers, compute the correlation of the pairs and plot them. Now, look at the pairs x_{i+2} and x_i. Compute their correlation.

1.3. Now modify your program from Exercise 1.2 so as to implement a matrix congruential method

$$\begin{bmatrix} x_{1i} \\ x_{2i} \end{bmatrix} \equiv \begin{bmatrix} a_{11} & a_{12} \\ a_{21} & a_{22} \end{bmatrix} \begin{bmatrix} x_{1,i-1} \\ x_{2,i-1} \end{bmatrix} \bmod m,$$

with $m = 2^{13} - 1$, $a_{11} = 17$, $a_{22} = 85$, and a_{12} and a_{21} variable. Letting a_{12} and a_{21} vary between 0 and 17, generate 500 vectors and compute their sample variance-covariances. Are the variances of the two elements in your vectors constant? Explain. What about the covariances? Can you see any relationship between the covariances and a_{12} and a_{21}? Is there any reason to vary a_{12} and a_{21} separately? Can a lower triangular matrix, that is, one with $a_{21} = 0$, provide all of the flexibility of matrices with varying values of a_{21}?

1.4. Write a Fortran or C function to implement the multiplicative congruential generator (1.3) (RANDU) on page 13.

   (a) Generate a sequence x_i of length 20,002. For all triplets in your sequence, (x_i, x_{i+1}, x_{i+2}), in which $0.5 \leq x_{i+1} \leq 0.51$, plot x_i versus x_{i+2}. Comment on the pattern of your scatterplot. (This is similar to the graphical analysis performed by Lewis and Orav, 1989.)

   (b) Generate a sequence of length 1002. Use a program that plots points in three dimensions and rotates the axes to rotate the points until the 15 planes are clearly visible. (A program that could be used for this is the S-Plus function spin, for example.)

1.5. Using an analysis similar to that leading to equation (1.4), determine the maximum number of different planes on which triplets from the generator (1.3) would lie if instead of 65 539, the multiplier were 65 541. Determine the number of different planes if the multiplier were 65 533. (Notice both of these multipliers are congruent to 5 mod 8, as James, 1990, suggested.)

1.6. Write a Fortran or C function to use a multiplicative congruential method with $m = 2^{31} - 1$ and $a = 16\,807$ (the "minimal standard").

1.7. Write a Fortran or C function to use a multiplicative congruential method with $m = 2^{31} - 1$ and $a = 950\,706\,376$. Test for correctness (not for statistical quality) by using a seed of 1 and generating 10 numbers.

1.8. Suppose a sequence is generated using a linear congruential generator with modulus m, beginning with the seed x_0. Show that this sequence and the sequence generated with the seed $m - x_0$ are antithetic. *Hint:* Use induction.

1.9. Suppose a sequence is generated by $x_{i+1} \equiv ax_i \bmod m$, and a second sequence is generated by $y_{i+1} \equiv by_i \bmod m$, where

$$b \equiv a^{c-1} \bmod m,$$

and c is the period. Prove that the sequences are in reverse order.

1.10. For the generator $x_i \equiv 16\,807 x_{i-1} \bmod (2^{31} - 1)$, determine the value x_0 that will yield the largest possible value for x_1. (This seed can be used as a test that the largest value yielded by the generator is less than 1. It is desirable to scale all numbers into the *open* interval $(0,1)$, because the numbers from the uniform generator may be used to an inverse CDF method for a distribution with an infinite range. To insure that this is the case, the value used for scaling must be greater than $2^{31} - 1$. See Gentle, 1990.)

1.11. In this exercise use a package that supports computations for number theory, such as Maple.

   (a) Check that the multipliers in Exercises 1.2, 1.5, and 1.10 are all primitive roots of the respective moduli.

   (b) Determine all of the primitive roots of the modulus $2^{13} - 1$ of Exercise 1.2.

1.12. Suppose for one stream from a given linear congruential generator the seed is 500, and for another stream from the same generator the seed is 1000. What is the approximate correlation between the two streams?

1.13. Consider the Wichmann/Hill random number generator (page 33). Because the moduli are relatively prime, the generator cannot yield an exact zero. Could a computer implementation of this generator yield a zero? First, consider a computer that has only two digits of precision. The answer is obvious. Now, consider a computer with a more realistic number system (such as whatever computer you use most often). How likely is the generator to yield a 0 on this computer? Perform some computations to explore the possibilities. Can you make a simple adjustment to the generator to prevent a 0 from occurring?

1.14. Prove the inequality occurring in the array of equations (1.17) on page 35. *Hint:* Define an appropriate random variable W, and then use the fact that $(E(W))^2 \le E(W^2)$.

1.15. Write pseudocode for a random number generator for a parallel processing computer. Be sure that your generator preserves independence of the separate streams. Also, consider the issues of the user interface (for example, what input does the user provide?)

Chapter 2

Transformations of Uniform Deviates: General Methods

Sampling of random variates from a nonuniform distribution is usually done by applying a transformation to uniform variates. Each realization of the nonuniform random variable might be obtained from a single uniform variate or from a sequence of uniforms. Some methods that use a sequence of uniforms require that the sequence be independent; other methods use a random walk sequence, a Markov chain.

For some distributions there may be many choices of algorithms for generation of random numbers. The algorithms differ in speed, in accuracy, in storage requirements, and in complexity of coding. Some of the faster methods are approximate; but given the current cost of computing, the speed-up resulting from an additional approximation beyond the approximation resulting from the ordinary finite-precision representation is not worth the accuracy loss. The methods we discuss in this chapter are all exact, so any approximation is a result of the ordinary rounding and truncation necessary in using a computer to simulate real numbers or infinite sets. Occasionally a research paper will contend that the quality of the random numbers generated by some particular method, such as Box-Muller or the ratio-of-uniforms, is bad, but the quality ultimately depends only on the quality of the underlying uniform generator. A particular method, however, may exacerbate some fault in the uniform generator.

After accuracy, the next most important criterion is speed. The speed of a random number generation algorithm has two aspects: the set-up time and the generation time. In most cases the generation time is the more important component to optimize. Whenever the set-up time is significant, the computer program can preserve the variables initialized, so that if the function is called again with the same parameters, the set-up step can be bypassed. In a case of relatively expensive set-up overhead, a software system may provide a second function for the same distribution with less set-up time.

The other two criteria mentioned above, storage requirements and complexity of coding, are generally of very little concern in selecting algorithms for production random number generators.

The methods discussed in this chapter are "universal" in the sense that they apply to almost any distribution. Some of these methods are better than others for a particular distribution or for a particular range of the distribution. These techniques are used, either singly or in combination, for particular distributions, as we see in Chapter 3.

Some of these methods, especially those that involve inverting a function, apply directly only to univariate random variables, whereas other methods apply immediately to multivariate random variables.

2.1   Inverse CDF Method

If X is a scalar random variable with a continuous cumulative distribution function (CDF) P_X, then the random variable

$$U = P_X(X)$$

has a $U(0, 1)$ distribution. (This is easy to show; you are asked to do that in Exercise 2.1, page 80.) This fact provides a very simple relationship with a uniform random variable U and a random variable X with distribution function P:

$$X = P_X^{-1}(U).$$

Use of this straightforward transformation is called the *inverse CDF* technique. The reason it works can be seen in Figure 2.1; over a range for which the derivative of the CDF (the density) is large, there is more probability of realizing a uniform deviate.

The inverse CDF relationship exists between any two continuous (nonsingular) random variables. If X is a continuous random variable with CDF P_X and Y is a continuous random variable with CDF P_Y, then

$$X = P_X^{-1}(P_Y(Y))$$

over the ranges of positive support. Use of this kind of relationship is a matching of "scores", that is, of percentile points, of one distribution with those of another distribution. In addition to the uniform, as above, this kind of transformation is sometimes used with the normal distribution.

Whenever the inverse of the distribution function is easy to compute, the inverse CDF method is a good one. It also has the advantage that basic relationships among a set of uniform deviates (such as order relationships) may result in similar relationships among the set of deviates from the other distribution.

Because it is relatively difficult to compute the inverse of some distribution functions of interest, however, the inverse CDF method is not as commonly used as its simplicity might suggest. Even when the inverse P^{-1} exists in closed form, evaluating it directly may be much slower than use of some alternative method

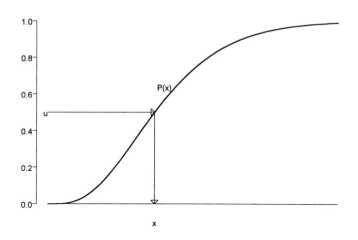

Figure 2.1: The Inverse CDF Method to Convert a Uniform Random Number to a Number from a Continuous Distribution

for sampling random numbers. On the other hand, in some cases when P^{-1} does not exist in closed form, use of the inverse CDF method by solving the equation

$$P(x) - u = 0$$

may be better than the use of any other method.

The inverse CDF method also applies to discrete distributions, but of course we cannot take the inverse of the distribution function. Suppose the discrete random variable X has mass points

$$m_1 < m_2 < m_3 < \ldots$$

with probabilities

$$p_1, \; p_2, \; p_3, \; \cdots,$$

and with the distribution function

$$P(x) = \sum_{i \ni m_i \leq x} p_i.$$

To use the inverse CDF method for this distribution, we first generate a realization u of the uniform random variable U. We then deliver the realization of the target distribution as x, where x satisfies the relationship

$$P(x_{(-)}) < u \leq P(x). \tag{2.1}$$

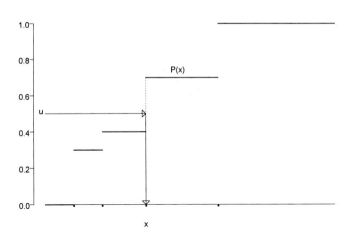

Figure 2.2: The Inverse CDF Method to Convert a Uniform Random Number to a Number from a Discrete Distribution

This is illustrated in Figure 2.2.

Without loss of generality, we often assume that the mass points of a discrete distribution are the integers $1, 2, 3, \ldots$. The special case in which there are k mass points and they all have equal probability is called the discrete uniform distribution, and the use of the inverse CDF method is particularly simple: the value is $\lceil uk \rceil$.

Using the inverse CDF method for a general discrete distribution is essentially a table lookup; it usually requires a search for the x in (2.1). The search may be performed sequentially or by some tree traversal. Marsaglia (1963), Norman and Cannon (1972), and Chen and Asau (1974) describe various ways of speeding up the table lookup. Improving the efficiency of the table-lookup method is often done by incorporating some aspects of the *urn method*, in which the distribution is simulated by a table (an "urn") that contains the mass points in proportion to their population frequency. In the urn each p_i is represented as a rational fraction n_i/N, and a table of length N is constructed with n_i pointers to the mass point i. A discrete uniform, $\lceil uN \rceil$, is then used as an index to the table to yield the target distribution.

The method of Marsaglia implemented by Norman and Cannon (1972), involves forming a table partitioned in such a way that the individual partitions can be sampled with equal probabilities. In this scheme, the probability p_i

associated with the ith mass point is expressed to a precision t in the base b as

$$p_i = \sum_{j=1}^{t} d_{ij} b^{-j},$$

with $0 \leq d_{ij} < b$. We assume there are n mass points. (We frequently assume a finite set of mass points when working with discrete distributions. Because we can work with only a finite number of different values on the computer, it does not directly limit our methods; but we should be aware of any such exclusion of rare events, and in some cases must modify our methods to be able to model rare events.) Let

$$
\begin{aligned}
P_0 &= 0 \\
P_j &= b^{-j} \sum_{i=1}^{n} d_{ij} \quad \text{for } j = 1, 2, \ldots, t \\
N_0 &= 0 \\
N_k &= b^{-j} \sum_{j=1}^{k} \sum_{i=1}^{n} d_{ij} \quad \text{for } k = 1, 2, \ldots, t.
\end{aligned}
$$

Now form a partitioned array and in the jth partition, store d_{ij} copies of i (remember the mass points are taken to be the integers; they could just as well be indexes). The jth partition is in locations $N_{j-1} + 1$ to N_j. There are N_t storage locations in all. Norman and Cannon (1972) show how to reduce the size of the table with a slight modification to the method. After the partitions are set up, Algorithm 2.1 generates a random number from the given distribution.

Algorithm 2.1 Marsaglia/Norman/Cannon Table Lookup for Sampling a Discrete Random Variate

1. Generate a uniform, u, and represent it in base b to t places:

$$u = \sum_{j=1}^{t} d_j b^{-j}.$$

2. Find m such that

$$\sum_{j=0}^{m-1} P_j \leq u < \sum_{j=0}^{m} P_j.$$

3. Take as the generated value the contents of location

$$\sum_{j=1}^{m} d_j b^{m-j} - \left(b^m \sum_{j=0}^{m-1} P_j - N_{m-1} \right) + 1.$$

■

In this algorithm the jth partition is chosen with probability P_j. The probability of the ith mass point is

$$\Pr(X = i) = \sum_{j=1}^{t} \Pr(j^{th} \text{ partition is chosen}) \times$$

$$\Pr(i \text{ is chosen from } j^{th} \text{ partition})$$

$$= \sum_{j=1}^{t} P_j \frac{d_{ij}}{\sum_{k=1}^{n} d_{kj}}$$

$$= \sum_{j=1}^{t} b^{-j} \sum_{i=1}^{n} d_{ij} \frac{d_{ij}}{\sum_{k=1}^{n} d_{kj}}$$

$$= \sum_{j=1}^{t} d_{ij} b^{-j}$$

$$= p_i.$$

Norman and Cannon (1972) give a program to implement this algorithm that formed an equivalent, but more compact partitioned array.

Chen and Asau (1974) give a hashing method using a "guide table". The guide table contains n values g_i that serve as indexes to the n mass points in the CDF table. The ith guide value is the index of the largest mass point whose CDF value is less than i/n:

$$g_i = \max_{\sum_{k=1}^{j} p_k < i/n} j.$$

After the guide table is set up, Algorithm 2.2 generates a random number from the given distribution.

Algorithm 2.2 Sampling a Discrete Random Variate Using the Chen and Asau Guide Table Method

1. Generate a uniform, u, set $i = \lceil un \rceil$.

2. Set $x = g_i + 1$.

3. While $\sum_{k=1}^{x} p_k > u$ set $x = x - 1$. ∎

Often for a continuous random variable we may have a table of values of the cumulative distribution function, but not have a function representing the CDF over its full range. This situation may arise in applications in which a person familiar with the process can assign probabilities for the variable of interest, yet may be unwilling to assume a particular distributional form. One approach to this problem is to fit a continuous function to the tabular values, and then use the inverse CDF method on the interpolant. The simplest interpolating function, of course, is the piecewise linear function; but second- or third-degree polynomials may give a better fit. It is important, however, that the interpolant be monotone. Guerra, Tapia, and Thompson (1976) describe a scheme for approximating the CDF based on an interpolation method of Akima (1970). Their procedure is implemented in the IMSL routine rngct.

An example of a common and very simple application of the inverse CDF technique is for generating a random deviate from a Bernoulli distribution with parameter π, as in Algorithm 2.3.

Algorithm 2.3 Generating a Bernoulli by the Inverse CDF

1. Generate a uniform, u.

2. If $u < \pi$, then
   2.a. deliver 0;
   otherwise
   2.b. deliver 1. ∎

The inverse CDF method does not apply to a multivariate distribution, although marginal and conditional univariate distributions can be used in an inverse CDF method to generate multivariate random variates. If the CDF of the multivariate random variable $(X_1, X_2, \ldots, X_d)$ is decomposed as

$$P_{X_1 X_2 \ldots X_d}(x_1, x_2, \ldots, x_d) =$$
$$P_{X_1}(x_1) P_{X_2|X_1}(x_2|x_1) \cdots P_{X_d|X_1 X_2 \ldots X_{d-1}}(x_d|x_1, x_2, \ldots, x_{d-1}),$$

and the functions are invertible, the inverse CDF method is applied sequentially, using independent realizations of a $U(0, 1)$ random variable, $u_1, u_2, \ldots, u_d$:

$$x_1 = P_{X_1}^{-1}(u_1)$$
$$x_2 = P_{X_2|X_1}^{-1}(u_2)$$
$$\cdots \quad \cdots$$
$$x_d = P_{X_d|X_1 X_2 \ldots X_{d-1}}^{-1}(u_d).$$

The modifications of the inverse CDF for discrete random variables described above can be applied if necessary.

2.2 Acceptance/Rejection Methods

For generating realizations of a random variable X, the *acceptance/rejection method* makes use of realizations of another random variable Y whose probability density g_Y is similar to the probability density of X, p_X. The random variable Y is chosen so that we can easily generate realizations of it and so that its density g_Y can be scaled to majorize p_X, using some constant c; that is, so that $cg_Y(x) \geq p_X(x)$ for all x. The density g_Y is called the *majorizing* density and cg_Y is called the majorizing function.

Unlike the inverse CDF method, the acceptance/rejection applies immediately to multivariate random variables.

Algorithm 2.4 The Acceptance/Rejection Method to Convert Uniform Random Numbers

1. Generate y from the distribution with density function g_Y.

2. Generate u from a uniform (0,1) distribution.

3. If $u \le p_X(y)/cg_Y(y)$, then
   3.a. take y as the desired realization;
   otherwise
   3.b. return to step 1.   ∎

It is easy to see that the random number delivered by Algorithm 2.4 has a density p_X. (In Exercise 2.2, page 80, you are asked to write the formal proof.)

Figure 2.3 illustrates the functions used in the acceptance/rejection method. (Actually, Figure 2.3 represents a specific case, namely, the beta distribution with parameters 3 and 2. In Exercise 2.3, page 81, you are asked to write a program implementing the acceptance/rejection method with the majorizing density shown.)

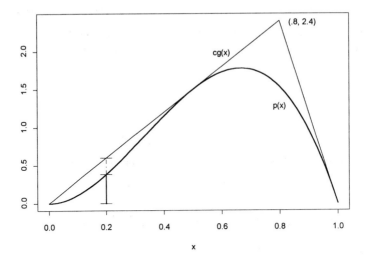

Figure 2.3: The Acceptance/Rejection Method to Convert Uniform Random Numbers

The acceptance/rejection method can be visualized as choosing a subsequence from a sequence of independently and identically distributed (i.i.d.) realizations from the distribution with density g_Y in such a way the subsequence has density p_X, as shown in Figure 2.4.

In the following, we generally dispense with the subscripts on the densities, except in some cases where we want to emphasize the random variable. It is important to remember, however, that the notation used for the argument of a function does not identify the function (despite the use of this convention by some statisticians).

Obviously, the closer $cg(x)$ is to $p(x)$, the faster the acceptance/rejection algorithm will be, if we ignore the time required to generate y from the dominating density g. A good majorizing function would be such that the l is almost as large as k in Figure 2.4. Often, g is chosen to be a very simple density, such as a uniform or a triangular density. When the dominating density is uniform, the acceptance/rejection method is similar to the "hit-or-miss" method described in Exercise 5.2 of Chapter 5, page 146.

There are many variations of the acceptance/rejection method. The method described here uses a sequence of i.i.d. variates from the majorizing density. It is also possible to use a sequence from a conditional majorizing density. A method using a nonindependent sequence is called a Metropolis method (and there are variations of these, with their own names, as we see below).

Acceptance/rejection methods, like any method for generating nonuniform random numbers, are dependent on a good source of uniforms. Hörmann and Derflinger (1993) illustrate that small values of the multiplier in a congruential generator for the uniforms can result in poor quality of the output from an acceptance/rejection method. Of course, we have seen that small multipliers are not good for generating uniforms. (See the discussion about Figure 1.2, page 10.) They rediscover the method of expression (1.6) and recommend using it so that larger multipliers can be used in the linear congruential generator.

Reducing the Computations in Acceptance/Rejection

Often, the target density, p, is difficult to evaluate, and so an easy way of speeding up the process is to use simple functions that bracket p to avoid the evaluation of p with a high probability. This method is called a "squeeze" (see Marsaglia, 1977). Most algorithms that use a squeeze function only use one below the density of interest. This allows quicker rejection. The squeeze function is often a linear or piecewise linear function.

Another procedure for making the acceptance/rejection decision with fewer computations is the "patchwork" method of Kemp (1990). In this method the unit square is divided into rectangles that correspond to pairs of uniforms

i.i.d. from g_Y	y_i	y_{i+1}	y_{i+2}	y_{i+3}	$\cdots$	y_{i+k}	$\cdots$
accept?	no	yes	no	yes	$\cdots$	yes	$\cdots$
i.i.d. from p_X		x_j		x_{j+1}	$\cdots$	x_{j+l}	$\cdots$

Figure 2.4: Acceptance/Rejection

that would lead to acceptance or to rejection or to lack of decision. The full evaluations for the acceptance/rejection algorithm need be performed only if the pair of uniforms to be used are in a rectangle of the latter type.

For a density that is nearly linear (or nearly linear over some range) Marsaglia (1962) describes some methods for efficient generation. Use of an inverse CDF method for a distribution whose density is exactly linear over some range involves a square root operation. Another simple way of generating from a linear density is to use the maximum order statistic of a sample of size two from a uniform distribution; that is, independently generate two $U(0,1)$ variates, u_1 and u_2, and use $\max(u_1, u_2)$. (Order statistics from a uniform distribution have a beta distribution; see Section 4.3.1, page 125.) Knuth (1981) gives a simple method for nearly linear densities. Following his development, suppose, as in Figure 2.5, the density over the interval $(s, s+h)$ is bounded by two parallel lines,

$$l_1(x) = a - b(x - s)/h$$

and

$$l_2(x) = b - b(x - s)/h.$$

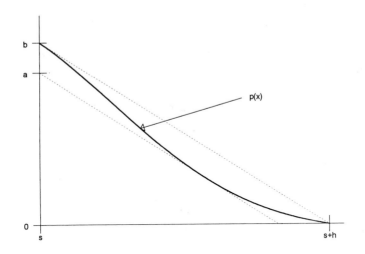

Figure 2.5: A Nearly Linear Density

Usually when we take advantage of the fact that a density is nearly linear, it is not the complete density that is linear, but rather the nearly linear density is combined with other densities to form the density of interest. (See the discussion of mixtures of densities in Section 2.3.) If, however, we assume that $p(x)$ in

Figure 2.5 is a density (that is, $\int_s^{s+h} p(x)\,dx = 1$), then Algorithm 2.5, which is Knuth's method, yields deviates from the distribution with density p. This algorithm implicitly uses a mixture of two distributions. Notice the use of the maximum of two uniforms, which we have mentioned earlier in connection with linear densities. (Also see Exercise 2.8, page 82.)

Algorithm 2.5 Sampling from a Nearly Linear Density

1. Generate u_1 and u_2 independently from a uniform (0,1) distribution. Set $u = \min(u_1, u_2)$, $v = \max(u_1, u_2)$, and $x = s + hu$.

2. If $v \le a/b$, then
   2.a. go to step 3;
   otherwise
   2.b. if $v > u + p(x)/b$, go to step 1.

3. Deliver x. ∎

Variations of Acceptance/Rejection

Wallace (1976) has introduced a modification of the acceptance/rejection method called *transformed rejection*. In the transformed acceptance/rejection method, the steps of Algorithm 2.4 are combined and rearranged slightly. Let G be the CDF corresponding to the dominating density g. Let $H(x) = G^{-1}(x)$, and let $h(x) = d\,H(x)/dx$. If v is a uniform (0,1) deviate, step 1 in Algorithm 2.4 is equivalent to $y = H(v)$, so we have Algorithm 2.6.

Algorithm 2.6 The Transformed Acceptance/Rejection Method

1. Generate u and v independently from a uniform (0,1) distribution.

2. If $u \le p(H(v))h(v)/c$, then
   2.a. take $H(v)$ as the desired realization;
   otherwise
   2.b. return to step 1. ∎

Marsaglia (1984) describes a method very similar to the transformed acceptance/rejection method: use ordinary acceptance/rejection to generate a variate x from the density proportional to $p(H(\cdot))h(\cdot)$ and then return $H(x)$. The choice of H is critical to the efficiency of the method of course. It should be close to the inverse of the CDF of the target distribution, P^{-1}. Marsaglia called this the *exact-approximation* method. Devroye (1986) calls the method *almost exact inversion*.

Acceptance/Rejection for Discrete Distributions

There are various ways that acceptance/rejection can be used for discrete distributions. One advantage of these methods is that they can be easily adapted to changes in the distribution. Rajasekaran and Ross (1993) consider the discrete random variable X_s such that

$$
\begin{aligned}
\Pr(X_s = x_i) &= p_{si} \\
&= \frac{a_{si}}{a_{s1} + a_{s2} + \cdots a_{sk}}, \quad i = 1, \ldots, k.
\end{aligned}
$$

(If $\sum_{i=1}^{k} a_{si} = 1$, the numerator a_{si} is the ordinary probability p_{si} at the mass point i.) Suppose that there exists an a_i^* such that $a_i^* \leq a_{si}$ for $s = 1, 2, \ldots$, and $b > 0$ such that $\sum_{i=1}^{k} a_{si} \geq b$ for $s = 1, 2, \ldots$. Let

$$
a^* = \max\{a_i^*\},
$$

and let

$$
P_{si} = a_{si}/a^* \quad \text{for } i = 1, \ldots, k.
$$

The generation method for X_s is shown in Algorithm 2.7.

Algorithm 2.7 Acceptance/Rejection Method for Discrete Distributions

1. Generate u from a uniform (0,1) distribution and let $i = \lceil ku \rceil$.

2. Let $r = i - ku$.

3. If $r \leq P_{si}$, then
   3.a. take i as the desired realization;
   otherwise
   3.b. return to step 1.   ∎

Suppose for the random variable X_{s+1}, $p_{s+1,i} \neq p_{si}$ for some i. (Of course, if this is the case for mass point i, it is also necessarily the case for some other mass point.) For each mass point whose probability changes, reset $P_{s+1,i}$ to $a_{s+1,i}/a^*$ and continue with Algorithm 2.7.

Rajasekaran and Ross (1993) also gave two other acceptance/rejection type algorithms for discrete distributions that are particularly efficient for use with distributions that may be changing. The other algorithms require slightly more preprocessing time, but yield faster generation times than Algorithm 2.7.

Other Applications of Acceptance/Rejection

The acceptance/rejection method can often be used to evaluate an elementary function at a random point. Suppose, for example, we wish to evaluate $\tan(\pi U)$ for U distributed as $U(-.5, .5)$. A realization of $\tan(\pi U)$ can be simulated by generating u_1 and u_2 independently from $U(-1, 1)$, checking if $u_1^2 + u_2^2 \leq 1$, and if so, delivering u_1/u_2 as $\tan(\pi u)$. (To see this, think of u_1 and u_2 as sine and cosine values.) Von Neumann (1951) gives an acceptance/rejection method for generating sines and cosines of random angles. An example of evaluating a logarithm can be constructed by use of the equivalence of an inverse CDF method and an acceptance/rejection method for sampling an exponential random deviate. (The methods are equivalent in a stochastic sense; they are both valid, but they will not yield the same stream of deviates.)

Acceptance/Rejection for Multivariate Distributions

The acceptance/rejection method is one of the most widely applicable methods for random number generation. It is used in many different forms, often in combination with other methods. It is clear from the description of the algorithm that the acceptance/rejection method applies equally to multivariate distributions. (The uniform random number is still univariate, of course.)

Example of Acceptance/Rejection: A Bivariate Gamma Distribution

Becker and Roux (1981)defined a bivariate extension of the gamma distribution that serves as a useful model for failure times for two related components in a system. (The model is also a generalization of a bivariate exponential distribution introduced by Freund, 1961; see Steel and Le Roux, 1987.) The probability density is given by

$$
p_{X_1 X_2}(x_1, x_2) = \begin{cases}
\begin{aligned}
&\lambda_2 \left(\Gamma(\alpha_1)\,\Gamma(\alpha_2)\,\beta_1^{\alpha_1}\beta_2^{\alpha_2}\right)^{-1} \times \\
&x_1^{\alpha_1-1} \left(\lambda_2(x_2 - x_1) + x_1\right)^{\alpha_2-1} \times \\
&\exp\left(-(\tfrac{1}{\beta_1} + \tfrac{1}{\beta_2} - \tfrac{\lambda_2}{\beta_2})x_1 - \tfrac{\lambda_2}{\beta_2}x_2\right) \quad \text{for} \quad 0 \le x_1 \le x_2,
\end{aligned} \\[2ex]
\begin{aligned}
&\lambda_1 \left(\Gamma(\alpha_1)\,\Gamma(\alpha_2)\,\beta_1^{\alpha_1}\beta_2^{\alpha_2}\right)^{-1} \times \\
&x_2^{\alpha_2-1} \left(\lambda_1(x_1 - x_2) + x_2\right)^{\alpha_1-1} \times \\
&\exp\left(-(\tfrac{1}{\beta_1} + \tfrac{1}{\beta_2} - \tfrac{\lambda_1}{\beta_1})x_2 - \tfrac{\lambda_1}{\beta_1}x_1\right) \quad \text{for} \quad 0 \le x_2 < x_1,
\end{aligned} \\[2ex]
0 \qquad\qquad\qquad\qquad\qquad\qquad\qquad \text{elsewhere.}
\end{cases}
$$
(2.2)

The density for $\alpha_1 = 4$, $\alpha_2 = 3$, $\beta_1 = 3$, $\beta_2 = 1$, $\lambda_1 = 3$, and $\lambda_2 = 2$ is shown in Figure 2.6.

It is a little more complicated to determine a majorizing density for this distribution. First of all, not many bivariate densities are familiar to us. The density must have support over the positive quadrant. A bivariate normal might be tried, but the $\exp(-(u_1 x_1 + u_2 x_2)^2)$ term in the normal density dies out more rapidly than the $\exp(-v_1 x_1 - v_2 x_2)$ term in the gamma. The normal cannot majorize the gamma in the limit.

We may be concerned about covariance of the variables in the bivariate gamma, but the fact that the variables have nonzero covariance is of little concern in using the acceptance/rejection method. The main thing, of course, is that we determine a majorizing density so that the probability of acceptance is high. We can use a bivariate density of independent variables as the majorizing

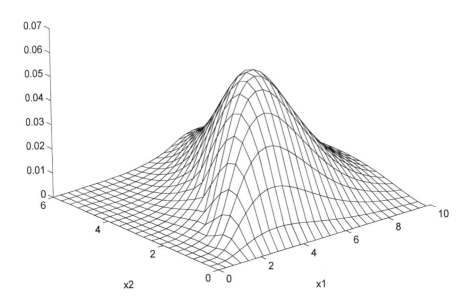

Figure 2.6: A Bivariate Gamma Density, Equation (2.2)

density. A product of two univariate densities would fit the bill nicely. A bivariate distribution of independent exponentials may work. Such a density has a maximum at $(0,0)$, however, and there would be a large volume between the bivariate gamma density and the majorizing function formed from a bivariate exponential. We can reduce this volume by choosing a bivariate uniform over the rectangle with corners $(0,0)$ and (z_1, z_2). Our majorizing density then is composed of two densities, a bivariate exponential,

$$
g_1(y_1, y_2) = \begin{cases} \frac{1}{v} \exp\left(-\frac{y_1}{\theta_1} - \frac{y_2}{\theta_2}\right) & \text{for} \quad \begin{array}{l} y_1 > z_1 \text{ and } y_2 > 0 \\ \text{or} \quad y_1 > 0 \text{ and } y_2 > z_2, \end{array} \\ 0 & \text{elsewhere,} \end{cases} \tag{2.3}
$$

where the constant v is chosen to make g_1 a density, and a bivariate uniform,

$$
g_2(y_1, y_2) = \begin{cases} \frac{1}{z_1 z_2} & \text{for} \quad 0 < y_1 \le z_1, \text{ and } 0 < y_2 \le z_2, \\ 0 & \text{elsewhere.} \end{cases} \tag{2.4}
$$

We next choose θ_1 and θ_2 so that the bivariate exponential can majorize the bivariate gamma. This requires that

$$
\frac{1}{\theta_1} \ge \max\left(\left(\frac{1}{\beta_1} + \frac{1}{\beta_2} - \frac{\lambda_2}{\beta_2}\right), \frac{\lambda_1}{\beta_1}\right),
$$

with a similar requirement for θ_2. Let us choose $\theta_1 = 1$ and $\theta_2 = 2$. Next, we choose z_1 and z_2 as the mode of the bivariate gamma. This point is $(4\frac{1}{3}, 2)$. We now choose c so that $cg_1(z_1, z_2) \geq p(z_1, z_2)$.

The method is

1. Generate a uniform u.

2. Generate (y_1, y_2) from a bivariate exponential such as (2.3), except over the full range, that is, with $v = \theta_1 \theta_2$.

3. If (y_1, y_2) is outside of the rectangle with corners $(0,0)$ and (z_1, z_2), then
    3.a. if $u \leq p(y_1, y_2)/cg_1(y_1, y_2)$ then
        3.a.i. deliver (y_1, y_2);
    otherwise
        3.a.ii. go to step 1.
  otherwise
    3.b. generate a new (y_1, y_2) as a bivariate uniform in that rectangle
        if $u \leq p(y_1, y_2)/(cy_1 y_2)$ then
        3.b.i. deliver (y_1, y_2);
    otherwise
        3.b.ii. go to step 1.

The majorizing density could be changed so that it is closer to the bivariate gamma. In particular, instead of the uniform over the rectangle with a corner on the origin, a pyramidal density that is closer to the bivariate gamma could be used.

2.3 Mixtures of Distributions

In the example of the bivariate gamma distribution, we used a majorizing density that corresponded to a truncated bivariate exponential over one range and a bivariate uniform over another range. It is often useful to break up the range of the distribution of interest in this way, using one density over one subrange and another density over another subrange. More generally, we may represent the distribution of interest as a mixture distribution. A mixture distribution is composed of proportions of other distributions. Suppose, for example, that the probability density or probability function of the random variable of interest, p, can be represented as

$$p(x) = w_1 p_1(x) + w_2 p_2(x), \qquad (2.5)$$

where p_1 and p_2 are density functions or probability functions of random variables, the union of whose support is the support of the random variable of interest. Obviously,

$$w_1 + w_2 = 1.$$

The random variable of interest has a *mixture distribution*.

To generate a random deviate from a mixture distribution, first use a single uniform to select the component distribution, and then generate a deviate from it. The mixture can consist of any number of terms. To generate a sample of n random deviates from a mixture distribution of d distributions, consider the proportions to be the parameters of a d-variate multinomial distribution. The first step is to generate a single multinomial deviate, then generate the required number of deviates from each of the component distributions.

Any decomposition of p into the sum of nonnegative functions yields the decomposition in equation (2.5), by choosing w_i so that the remaining function integrates to 1.

For example, suppose a distribution has density $p(x)$, and for some constant c, $p(x) \geq c$ over (a, b). Then the distribution can be decomposed into a mixture of a uniform over (a, b) with proportion $c(b-a)$ and some leftover part, say $g(x)$. Now $g(x)/(1 - c(b-a))$ is a probability density function. To generate a deviate from p:

>with probability $c(b - a)$,
>>generate a uniform (a, b);
>otherwise
>>generate from the density $g(x)/(1 - c(b - a))$.

If $c(b - a)$ is close to 1, we will generate from the uniform distribution most of the time, so even if it is difficult to generate from $g(x)/(1 - c(b - a))$, this decomposition of the original distribution may be useful.

Another way of forming a mixture distribution is to consider a density similar to (2.5) that is a conditional density,

$$p(x|y) = yp_1(x) + (1 - y)p_2(x),$$

where y is the realization of a Bernoulli random variable, Y. If Y takes a value of 0 with probability $w_1/(w_1 + w_2)$, then the density in (2.5) is the marginal density. This conditional distribution yields

$$
\begin{aligned}
p_X(x) &= \int p_{X,Y}(x, y)\mathrm{d}y \\
&= \sum_y p_{X|Y=y}\Pr(Y = y) \\
&= w_1 p_1(x) + w_2 p_2(x),
\end{aligned}
$$

as in equation (2.5).

More generally, for any random variable X whose distribution is parameterized by θ, we can think of the parameter as being the realization of a random variable Θ. Some common distributions result from mixing other distributions; for example, if the gamma distribution is used to generate the parameter in a Poisson distribution, a negative binomial distribution is formed. Mixture distributions are often useful in their own right; for example, the beta-binomial distribution, which can be used to model overdispersion (see Ahn and Chen, 1995).

2.4 Mixtures and Acceptance Methods

Suppose, as above, the density of interest, p, may be written as

$$p(x) = w_1 p_1(x) + w_2 p_2(x),$$

and suppose there is a density g that majorizes $w_1 p_1$, that is, $g(x) \geq w_1 p_1(x)$ for all x.

Kronmal and Peterson (1981, 1984) consider this case and propose the following algorithm, which they call the acceptance/complement method.

Algorithm 2.8 The Acceptance/Complement Method to Convert Uniform Random Numbers

1. Generate y from the distribution with density function g.

2. Generate u from a uniform $(0,1)$ distribution.

3. If $u > w_1 p_1(y)/g(y)$, then generate y from the density p_2.

4. Take y as the desired realization. ∎

We discussed nearly linear densities and gave Knuth's algorithm for generating from such densities as Algorithm 2.5. Devroye (1986) gives an algorithm for a special nearly linear density, namely, one that is almost flat. The method is based on a simple decomposition using the supremum of the density. (In practice, as we have indicated in discussing other techniques, this method would probably be used for a component of a density that has already been decomposed.) To keep the description simple, assume that the range of the random variable is $(-1, 1)$ and the density p satisfies

$$\sup_x p(x) - \inf_x p(x) \leq \frac{1}{2}$$

over that interval. Now because p is a density, we have

$$0 \leq \inf_x p(x) \leq \frac{1}{2} \leq \sup_x p(x)$$

and

$$\sup_x p(x) \leq 1.$$

Let $p^* = \sup_x p(x)$ and decompose the target density into

$$p_1(x) = p(x) - (p^* - \frac{1}{2})$$

and

$$p_2(x) = (p^* - \frac{1}{2})$$

The method is shown in Algorithm 2.9.

Algorithm 2.9 Sampling from a Nearly Flat Density

1. Generate u from $U(0,1)$.

2. Generate x from $U(-1,1)$.

3. If $u > 2(p(x) - (p^* - \frac{1}{2}))$, then generate x from $U(-1,1)$.

4. Deliver x. ∎

Another variation on the general theme of acceptance/rejection applied to mixtures was proposed by Deák (1981) in what he called the "economical method". To generate a deviate from the density p using this method, an auxiliary density g is used, and an "excess area" and a "shortage area" are defined. The excess area is where $g(x) > p(x)$, and the shortage area is where $g(x) \leq p(x)$. We define two functions p_1 and p_2:

$$
\begin{aligned}
p_1(x) &= g(x) - p(x), \quad \text{if} \quad g(x) - p(x) < 0, \\
&= 0, \quad \text{otherwise,}
\end{aligned}
$$

$$
\begin{aligned}
p_2(x) &= p(x) - g(x), \quad \text{if} \quad p(x) - g(x) \geq 0, \\
&= 0, \quad \text{otherwise.}
\end{aligned}
$$

Now we define a transformation T that will map the excess area into the shortage area in a way that will yield the density p. Such a T is not unique, but one transformation that will work is

$$
T(x) = \min \left\{ t; \int_{-\infty}^{x} p_1(s) \, ds = \int_{-\infty}^{t} p_2(s) \, ds \right\}.
$$

Algorithm 2.10 shows the method.

Algorithm 2.10 The Economical Method to Convert Uniform Random Numbers

1. Generate y from the distribution with density function g.

2. If $p(y)/g(y) < 1$, then
   2.a. generate u from a uniform $(0,1)$ distribution
   2.b. if $u \leq p(y)/g(y)$, then replace y with $T(y)$.

3. Take y as the desired realization. ∎

Using the representation of a discrete distribution that has k mass points as an equally weighted mixture of k two-point distributions, Deák (1986) develops a version of the economical method for discrete distributions. (See Section 2.6, page 61, on the alias method for additional discussion of two-point representations.)

2.5 Ratio-of-Uniforms Method

Kinderman and Monahan (1977) discuss a very useful relationship among random variables U, V, and V/U:

If (U, V) is uniformly distributed over the set

$$C = \left\{ (u, v); \ 0 \le u \le \sqrt{h\left(\frac{v}{u}\right)} \right\},$$

where h is a nonnegative integrable function, then V/U has probability density proportional to h.

Use of this relationship is called a *ratio-of-uniforms* method. In practice, we may choose a rectangle R that encloses C, generate a uniform point in the rectangle, and reject a point that does not satisfy

$$u \le \sqrt{h\left(\frac{v}{u}\right)}.$$

If $h(x)$ and $x^2 h(x)$ are bounded in C, a simple form of the rectangle R is

$$\{(u, v); \ 0 \le u \le b, \ c \le v \le d\},$$

where

$$
\begin{aligned}
b &= \sup_x \sqrt{h(x)} \\
c &= \inf_x x\sqrt{h(x)} \\
d &= \sup_x x\sqrt{h(x)}.
\end{aligned}
$$

This yields the method shown in Algorithm 2.11.

Algorithm 2.11 Ratio-of-Uniforms Method (for Continuous Variates)

1. Generate u and v independently from a uniform (0,1) distribution.

2. Set $u_1 = bu$, and $v_1 = c + (d - c)v$.

3. Set $x = v_1/u_1$.

4. If $u_1^2 \le h(x)$, then
     4.a. take x as the desired realization;
   otherwise
     4.b. return to step 1.

The ratio-of-uniforms method is very simple to apply, and it can be quite fast. Cheng and Feast (1979) use the method for the gamma distribution, and Kinderman and Monahan (1980) apply it to the t distribution.

Wakefield, Gelfand, and Smith (1991) generalize the ratio-of-uniforms method by introducing a strictly increasing, differentiable function g that has the property $g(0) = 0$:

If (u, v) is uniformly distributed over the set

$$C_{h,g} = \left\{ (u, v); \ 0 \le u \le g \left(ch \left(\frac{v}{g'(u)} \right) \right) \right\},$$

where c is a positive constant and h is a nonnegative integrable function as before, then $v/g'(u)$ has probability density proportional to h.

Wakefield et al. also applied the ratio-of-uniforms method to some multivariate distributions.

Stadlober (1990, 1991) clarified the relationship of the ratio-of-uniforms method to the ordinary acceptance/rejection method, and applied the ratio-of-uniforms method to discrete distributions. If (U, V) is uniformly distributed over the rectangle

$$\{(u, v); \ 0 \le u \le 1, \ -1 \le v \le 1\},$$

and $X = sV/U + a$, for any $s > 0$, then X has the density

$$g_X(x) = \begin{cases} \dfrac{1}{4s}, & a - s \le x \le a + s, \\[2ex] \dfrac{s}{4(x-a)^2} & \text{elsewhere,} \end{cases}$$

and the conditional density of $Y = U^2$, given X is

$$g_{Y|X}(y|x) = \begin{cases} 1 & \text{for} \quad a - s \le x \le a + s, \quad \text{and} \quad 0 \le y \le 1, \\[2ex] \dfrac{(x-a)^2}{s^2} & \text{for} \quad x > |a + s|, \qquad \text{and} \quad 0 \le y \le \dfrac{s^2}{(x-a)^2}, \\[2ex] 0 & \text{elsewhere.} \end{cases}$$

The conditional distribution of Y given $X = x$ is uniform on $(0, 4sg(x))$, and the ratio-of-uniforms method is an acceptance/rejection method with a table mountain majorizing function.

Stadlober (1990) gives the modification of the ratio-of-uniforms in Algorithm 2.12 for a general discrete random variable with probability function $p(\cdot)$.

Algorithm 2.12 Ratio-of-Uniforms Method for Discrete Variates

1. Generate u and v independently from a uniform $(0,1)$ distribution.

2. Set $x = \lfloor a + s(2v - 1)/u \rfloor$.

3. Set $y = u^2$.

4. If $y \le p(x)$, then

    4.a. take x as the desired realization;

    otherwise

        4.b. return to step 1. ∎

Ahrens and Dieter (1991) describe a ratio-of-uniforms algorithm for the Poisson distribution, and Stadlober (1991) describes one for the binomial distribution.

Afflerbach and Hörmann (1992) and Hörmann (1994b) indicate that in some cases, the quality of the pseudorandom output stream from a ratio-of-uniforms method can be quite poor. The ratio-of-uniforms method transforms all points lying on one line through the origin into a single number. Because of the lattice structure of the uniforms from a linear congruential generator, the lines passing through the origin have regular patterns, which result in structural gaps in the numbers yielded by the ratio-of-uniforms method. Noting these distribution problems, Hörmann and Derflinger (1994) make some comparisons of the ratio-of-uniforms method with the transformed rejection method (Algorithm 2.6, page 51); and based on their empirical study, they recommend the transformed rejection method over the ratio-of-uniforms method. The quality of the output of the ratio-of-uniforms method, however, is more a function of the quality of the uniform generator, and would usually not be of any concern if a good uniform generator is used. The relative computational efficiencies of the two methods depend on the majorizing functions used.

2.6 Alias Method

Walker (1977) shows that a discrete distribution with k mass points can be represented as an equally weighted mixture of k two-point distributions, that is, distributions with only two mass points. Consider the random variable X such that

$$\Pr(X = x_i) = p_i, \quad i = 1, \ldots, k,$$

and $\sum_{i=1}^{k} p_i = 1$. Walker constructed k two-point distributions,

$$\Pr(Y_i = y_{ij}) = q_{ij}, \quad j = 1, 2; \quad i = 1, \ldots, k,$$

(with $q_{i1} + q_{i2} = 1$) in such a way that any p_i can be represented as k^{-1} times a sum of $q_{i,j}$'s. (It is easy to prove that this can be done; use induction, starting with $k = 1$.)

A setup procedure for the alias method is shown in Algorithm 2.13. The setup phase associates with each $i = 1$ to k a value P_i that will determine whether the original mass point or an "alias" mass point, indexed by a_i, will be delivered when i is chosen with equal probability, $\frac{1}{k}$. Two lists, L and H, are maintained to determine which points or point-pairs have probabilities less than or greater than $\frac{1}{k}$. At termination of the setup phase all points or point-pairs have probabilities equal to $\frac{1}{k}$. Marsaglia calls the setup phase "leveling the histogram". The outputs of the setup phase are two lists, P and a, each of length k.

Algorithm 2.13 Alias Method Setup to Initialize the Lists a and P

0. For $i = 1$ to k,
   set $a_i = i$;
   set $P_i = 0$;
   set $b_i = p_i - \frac{1}{k}$;
   and if $b_i < 0$, put i in the list L;
   otherwise put i in the list H.

1. If $\max(b_i) = 0$, stop.

2. Select $l \in L$ and $h \in H$.

3. Set $c = b_l$ and set $d = b_h$.

4. Set $b_l = 0$ and set $b_h = c + d$.

5. Remove l from L.

6. If $b_h \leq 0$, remove h from H; and if $b_h < 0$, put h in L.

7. Set $a_l = h$ and set $P_l = 1 + kc$.

8. Go to step 1.   ∎

Notice that $\sum b_i = 0$ during every step. The steps are illustrated in Figure 2.7 for a distribution such that

$$\begin{aligned}
\Pr(X = 1) &= .30 \\
\Pr(X = 2) &= .05 \\
\Pr(X = 3) &= .20 \\
\Pr(X = 4) &= .40 \\
\Pr(X = 5) &= .05
\end{aligned}$$

At the beginning $L = \{2, 5\}$ and $H = \{1, 4\}$. In the first step, the values corresponding to 2 and 4 are adjusted.

The steps to generate deviates, after the values of P_i and a_i are computed by the setup, are shown in Algorithm 2.14.

Algorithm 2.14 Generation Using the Alias Method, Following the Setup in Algorithm 2.13

1. Generate u from a uniform $(0,1)$ distribution.

2. Generate i from a discrete uniform over $1, 2, \ldots, k$.

3. If $u \leq P_i$, then
   3.a. deliver x_i;
   otherwise
   3.b. deliver x_{a_i}.   ∎

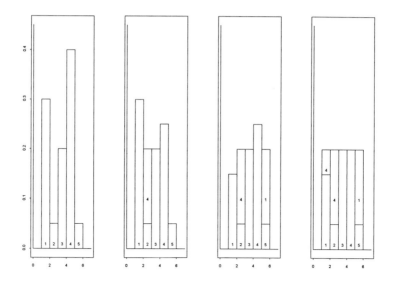

Figure 2.7: Setup for the Alias Method; Leveling the Histogram

It is clear that the setup time for Algorithm 2.13 is $O(k)$, because the total number of items in the lists L and H goes down by at least one at each step. If in step 2, the minimum and maximum values of b are found, as in the original algorithm of Walker (1977), the algorithm may proceed slightly faster in some cases, but then the algorithm is $O(k \log k)$. The method given in Algorithm 2.13 is from Kronmal and Peterson (1979a). Vose (1991) also describes a setup procedure that is $O(k)$. Once the setup is finished by whatever method, the generation time is constant, or $O(1)$.

In the IMSL Libraries, the routine **rngda** performs both the setup and the generation of discrete random deviates using an alias method.

Kronmal and Peterson (1979a, 1979b) apply the alias method to mixture methods and acceptance/rejection methods for continuous random variables. Peterson and Kronmal (1982) describe a modification of the alias method incorporating some aspects of the urn method. This hybrid method, which they called the *alias-urn* method, reduces the burden of comparisons at the expense of slightly more storage space.

2.7 Use of Stationary Distributions of Markov Chains

If the density of interest, p, is the density of the stationary distribution of a Markov chain, correlated samples from the distribution can be generated by simulating the Markov chain. (A *Markov chain* is a sequence of random variables, $X_1, X_2, \ldots$, such that the disribution of X_{t+1} given X_t is independent of $X_{t-1}, X_{t-2}, \ldots$. A sequence of realizations of such random variables is also called a Markov chain. An aperiodic, irreducible, positive recurrent Markov chain is associated with a *stationary distribution* or *invariant distribution*, which is the limiting distribution of the chain. See Meyn and Tweedie, 1993, or Tierney, 1996, for relevant definitions and discussions of properties of Markov chains.) An algorithm based on a stationary distribution of a Markov chain is an *iterative method* because a sequence of operations must be performed until they *converge*.

A Markov chain is the basis for several schemes for generating random numbers. The interest is not in the sequence of the Markov chain itself. The elements of the chain are accepted or rejected in such a way as to form a different chain whose stationary distribution is the distribution of interest.

Following engineering terminology for sampling sequences, the techniques based on these chains are generally called "samplers". The static sample, and not the sequence, is what is used. The objective in the Markov chain samplers is to generate a sequence of autocorrelated points with a given stationary distribution.

For a distribution with density p, the Metropolis algorithm, introduced by Metropolis et al. (1953) generates a random walk and performs an acceptance/rejection based on p evaluated at successive steps in the walk. In the simplest version, the walk moves from the point y_i to a candidate point $y_{i+1} = y_i + s$, where s is a realization from $U(-a, a)$, and accepts y_{i+1} if

$$\frac{p(y_{i+1})}{p(y_i)} \geq u,$$

where u is an independent realization from $U(0, 1)$. This method is also called the "heat bath" method because of the context in which it was introduced. The random walk of Metropolis et al. is the basic algorithm of *simulated annealing*, which is currently widely used in optimization problems.

If the range of the distribution is finite, the random walk is not allowed to go outside of the range. Consider, for example, the von Mises distribution, with density,

$$p(x) = \frac{1}{2\pi I_0(c)} e^{c \cos(x)}, \quad \text{for } -\pi \leq x \leq \pi, \tag{2.6}$$

where I_0 is the modified Bessel function of the first kind and of order zero. Notice, however, that it is not necessary to know this normalizing constant, because it is canceled in the ratio. The fact that all we need is a nonnegative function that is proportional to the density of interest is an important property of this method.

If $c = 3$, after a quick inspection of the amount of fluctuation in p, we may choose $a = 1$. The following Matlab statements implement the Metropolis algorithm to generate $n - 1$ deviates from the von Mises distribution.

```
while i < n
  yip1 = yi + 2*a*rand - 1;
  if yip1 < pi & yip1 > - pi
    i = i+1;
    if exp(c*(cos(yip1)-cos(yi))) > rand
       yi = yip1;
    else
       yi = x(i-1);
    end
    x(i) = yip1;
  end
end
```

The output for $n = 1000$ and a starting value of $y_0 = 1$ is shown in Figure 2.8. The output is a Markov chain. A histogram, which is not affected by the

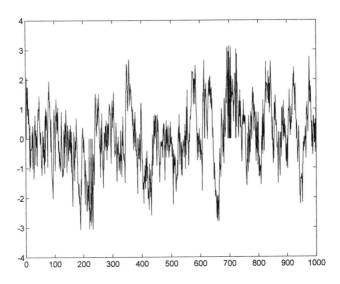

Figure 2.8: Sequential Output from the Metropolis Algorithm for a Von Mises Distribution

sequence of the output in a large sample, is shown in Figure 2.9.

The von Mises distribution is an easy one to simulate by the Metropolis algorithm. This distribution is often used by physicists in simulations of lattice gauge and spin models, and the Metropolis method is widely used in these simulations. Notice the simplicity of the algorithm: we did not need to determine a majorizing density, nor even evaluate the Bessel function that is the normalizing constant for the von Mises density.

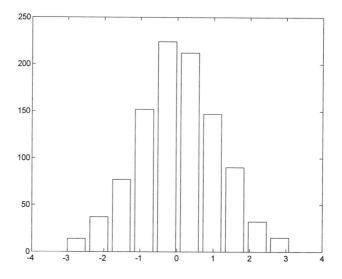

Figure 2.9: Histogram of the Output from the Metropolis Algorithm for a Von Mises Distribution

The Markov chain samplers require a "burn-in" period, that is, a number of iterations before the stationary distribution is achieved. In practice, the variates generated during the burn-in period are discarded. The number of iterations needed varies with the distribution, and can be quite large, sometimes several hundred. The von Mises example shown in Figure 2.8 is unusual; no burn-in is required. In general, convergence is much quicker for univariate distributions with finite ranges such as this one.

It is important to remember what convergence means; it does *not* mean that the sequence is independent from the point of convergence forward. The deviates are still from a Markov chain.

The Metropolis acceptance/rejection sequence is illustrated in Figure 2.10. Compare this with the acceptance/rejection method based on independent variables, as illustrated in Figure 2.13.

random walk	y_i	$y_{i+1} =$	$y_{i+3} =$	$y_{i+2} =$	
		$y_i + s_{i+1}$	$y_{i+1} + s_{i+2}$	$y_{i+2} + s_{i+3}$	$\cdots$
accept?	no	yes	no	yes	$\cdots$
i.i.d. from p_X		x_j		x_{j+1}	$\cdots$

Figure 2.10: Metropolis Acceptance/Rejection

Hastings (1970) developed an algorithm that uses a more general chain for the acceptance/rejection step. The *Metropolis-Hastings sampler* to generate deviates from a distribution with density p_X uses deviates from a Markov chain with density $g_{Y_{t+1}|Y_t}$. The method is shown in Algorithm 2.15. The conditional density $g_{Y_{t+1}|Y_t}$ is chosen so that it is easy to generate deviates from it.

Algorithm 2.15 Metropolis-Hastings Algorithm

0. Set $k = 0$.

1. Choose $x^{(k)}$ in the range of p_X. (The choice can be arbitrary.)

2. Generate y from the density $g_{Y_{t+1}|Y_t}(y|x^{(k)})$.

3. Set r:

$$r = p_X(y) \frac{g_{Y_{t+1}|Y_t}(x^{(k)}|y)}{p_X(x^{(k)}) g_{Y_{t+1}|Y_t}(y|x^{(k)})}$$

4. If $r \geq 1$, then
   4.a. set $x^{(k+1)} = y$;
   otherwise
      4.b. generate u from uniform(0,1) and
         if $u < r$, then
            4.b.i. set $x^{(k+1)} = y$,
         otherwise
            4.b.ii. set $x^{(k+1)} = x^{(k)}$.

5. If convergence has occurred, then
      5.a. deliver $x = x^{(k+1)}$;
   otherwise
      5.b. set $k = k + 1$, and go to step 2.   ∎

Compare Algorithm 2.15 with the basic acceptance/rejection method in Algorithm 2.4, page 48. The majorizing function in the Metropolis-Hastings algorithm is

$$\frac{g_{Y_{t+1}|Y_t}(x|y)}{p_X(x) g_{Y_{t+1}|Y_t}(y|x)}.$$

In Algorithm 2.15, r is called the "Hastings ratio", and step 4 is called the "Metropolis rejection". The conditional density, $g_{Y_{t+1}|Y_t}(\cdot|\cdot)$ is called the "proposal density" or the "candidate generating density". Notice that because the majorizing function contains p_X as a factor, we only need to know p_X to within a constant of proportionality. As we have mentioned already, this is an important characteristic of the Metropolis algorithms.

As with the acceptance/rejection methods with independent sequences, the acceptance/rejection methods based on Markov chains apply immediately to multivariate random variables.

We can see that this algorithm delivers realizations from the density p_X by using the same method suggested in Exercise 2.2 (page 80), that is, determine the CDF and differentiate. The CDF is the probability-weighted sum of the two components corresponding to whether the chain moved or not. In the case in which the chain does move, that is, in the case of acceptance, for the random variable Z whose realization is y in Algorithm 2.15, we have

$$
\begin{aligned}
\Pr(Z \leq x) &= \Pr\left(Y \leq x \,\middle|\, U \leq p(Y)\frac{g(x_i|Y)}{p(x_i)g(Y|x_i)}\right) \\
&= \frac{\int_{-\infty}^{x}\int_{0}^{p(t)g(x_i|t)/(p(x_i)g(t|x_i))} g(t|x_i)\, ds\, dt}{\int_{-\infty}^{\infty}\int_{0}^{p(t)g(x_i|t)/(p(x_i)g(t|x_i))} g(t|x_i)\, ds\, dt} \\
&= \int_{-\infty}^{x} p_X(t)\, dt.
\end{aligned}
$$

We can illustrate the use of the Metropolis-Hastings algorithm using a Markov chain in which the density of X_{t+1} is normal with a mean of X_t and a variance of σ^2. Let us use this density to generate a sample from a standard normal distribution (that is, a normal with a mean of 0 and a variance of 1). We start with a x_0, chosen arbitrarily. We take logs and cancel terms in the expression for r in Algorithm 2.15. The following simple Matlab statements generate the sample (and plot the sequence):

```
x(1) = x0;
while i < n
  i = i + 1;
  yip1 = yi + sigma*randn;
  lr2 = yi^2 - yip1^2;
  if lr2 > 0
    yi = yip1;
  else
    u = rand;
    if lr2 > log(u)*2
      yi = yip1;
    else
      yi = x(i-1);
    end
  end
  x(i) = yi;
end
plot (x)
```

The sequential output for $n = 1000$, a starting value of $x_0 = 10$ (yi = 10), and a variance of $\sigma^2 = 9$ (sigma = 3) is shown in Figure 2.11. Notice that the values descend very quickly from the starting value, which would be a very unusual realization of a standard normal. In practice, we generally cannot expect such a short burn-in period. Notice also in Figure 2.11 the horizontal line segments where the underlying Markov chain did not advance.

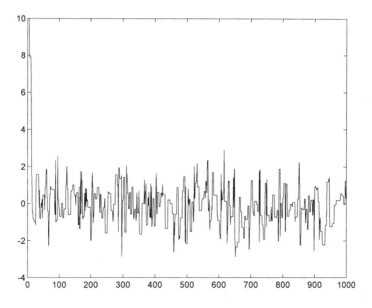

Figure 2.11: Sequential Output from a Standard Normal Distribution Using a Markov Chain, $N(X_t, \sigma^2)$

There are several variations of the basic Metropolis-Hastings algorithm. See Bhanot (1988)and Chib and Greenberg (1995) for descriptions of modifications and generalizations. Also see Section 2.10 for two related methods. These algorithms are used extensively in a class of methods called *Markov chain Monte Carlo*, or *MCMC*.

Markov chain Monte Carlo has become one of the most important tools in statistics in recent years. Its applications pervade Bayesian analysis, as well as many Monte Carlo procedures in the frequentist approach to statistical analysis. See Gilks, Richardson, and Spiegelhalter (1996) for several examples.

Whenever a correlated sequence such as a Markov chain is used, variance estimation must be performed with some care. In the more common cases of positive autocorrelation, the ordinary variance estimators are negatively biased. The method of batch means or some other method that attempts to account for the autocorrelation should be used. See Section 5.2 for discussions of these methods.

Tierney (1991) and (1994) describes an *independence sampler*, which is a Metropolis-Hasting sampler whose proposal density does not depend on Y_t: $g_{Y_{t+1}|Y_t}(\cdot|\cdot) = g_{Y_{t+1}}(\cdot)$. For this type of proposal density, it is more critical that $g_{Y_{t+1}}(\cdot)$ approximates $p_X(\cdot)$ fairly well, and that it can be scaled to majorize $p_X(\cdot)$ in the tails. Liu (1996) and Roberts (1996) discuss some of the properties of the independence sampler and its relationship to other Metropolis-Hastings methods.

Some of the most important issues in MCMC concern the rate of convergence, that is, the length of the burn-in, and the frequency with which the chain advances. In most applications of simulation, such as studies of waiting times in queues, there is more interest in transient behavior than in stationary behavior. This is not the case in random number generation using an iterative method. For general use in random number generation, the stationary distribution is the only thing of interest. (We often use the terms "Monte Carlo" and "simulation" rather synonymously; stationarity and transience, however, are often the key distinction between Monte Carlo applications and simulation applications. In simulation in practice, the interest is rarely in the stationary behavior, but it is in these Monte Carlo applications.)

The issue of convergence is more difficult to address in multivariate distributions. It is for multivariate distributions, however, that the MCMC method is most useful. This is because the Metropolis-Hastings algorithm does not require knowledge of the normalizing constants, and the computation of a normalizing constant may be more difficult for multivariate distributions.

To observe the performance of MCMC in higher dimensions, consider an example similar to that shown in Figure 2.11, except for a multivariate normal distribution instead of a univariate one. We use a d-dimensional normal with a mean vector x_t and a variance-covariance matrix Σ to generate x_{t+1} for use in the Metropolis-Hastings method of Algorithm 2.15. Taking $d = 3$,

$$\Sigma = \begin{bmatrix} 9 & 0 & 0 \\ 0 & 9 & 0 \\ 0 & 0 & 9 \end{bmatrix},$$

and starting with $x_0 = (10, 10, 10)$, the first 1000 values of the first element (which should be a realization from a standard univariate normal) are shown in Figure 2.12.

The example shown was generated by the following simple Matlab statements:

```
x(1,:) = x0;
while i < n
  i = i + 1;
  yip1 = yi + (half*randn(size(yi)))')';
  lr2 = yi'*yi - yip1'*yip1;
  if lr2 > 0
    yi = yip1;
  else
    u = rand;
    if lr2 > log(u)*2
      yi = yip1;
    else
      yi = x(i-1,:);
    end
```

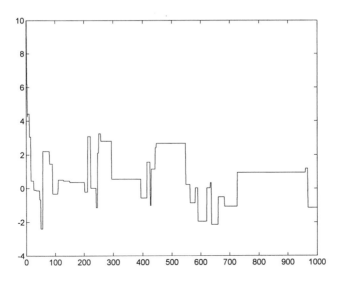

Figure 2.12: Sequential Output of x_1 from a Trivariate Standard Normal Distribution Using a Markov Chain, $N(X_t, \Sigma)$

```
    end
    x(i,:) = yi;
end
plot (x(:,1))
```

Gelman and Rubin (1992b) give examples in which the burn-in is much longer than might be expected. Various diagnostics have been proposed to assess convergence. Cowles and Carlin (1996) discuss and compare thirteen different ones. Most of these diagnostics use multiple chains in one way or another; see, for example, Gelman and Rubin (1992a), Johnson (1996), and Roberts (1992). Use of batch means from separate streams can be used to determine when the variance has stabilized. (See Section 5.2 for a description of batch means.) Yu (1995) uses a cusum plot on only one chain to help to identify convergence. Rosenthal (1995) provides some general methods for placing bounds on the length of runs required to give satisfactory results. All of these methods have limitations. Various methods have been proposed to speed up the convergence; see Gelfand and Sahu (1994), for example. Frigessi, Martinelli, and Stander (1997) discuss general issues of convergence and acceleration of convergence.

For a monotone chain (one whose transition matrix stochastically preserves orderings of state vectors) that has two starting state vectors x_0^- and x_0^+ such that for all $x \in S$, $x_0^- \leq x \leq x_0^+$, Propp and Wilson (1996) show that if the sequence beginning with x_0^- and the sequence beginning with x_0^+ coalesce, then the sequence from that point on is a sample from the stationary distribution. Because once the sequences have coalesced, the continuing sequence is known to be from the stationary distribution, Propp and Wilson call the algorithm *exact sampling*.

Methods of assessing convergence is currently an area of active research. The collection of articles in Gilks, Richardson, and Spiegelhalter (1996) addresses many of the problems of convergence. Gamerman (1997) provides a general introduction to MCMC in which many of these convergence issues are explored.

2.8   Weighted Resampling

To obtain a sample $x_1, x_2, \ldots, x_m$ that has an approximate distribution with density p_X, a sample $y_1, y_2, \ldots, y_n$ from another distribution with density g_Y can be resampled using weights, or probabilities,

$$w_i = \frac{p_X(y_i)/g_Y(x_i)}{\sum_{j=1}^{n} p_X(y_j)/g_Y(x_j)}, \quad \text{for } i = 1, 2, \ldots n.$$

The method was suggested by Rubin (1987, 1988), who called the method SIR for *sampling/importance resampling*. The method is also called importance-weighted resampling. The resampling should be done without replacement so as to give points with low probabilities a chance to be represented. Methods for sampling from a given set with given probabilities are discussed in Section 4.1, page 121. Generally in SIR n is much larger than m. This method can work reasonably well if the density g_Y is very close to the target density p_X.

This method, like the Markov chain methods above, has the advantage that the normalizing constant of the target density is not needed. Instead of the density $p_X(\cdot)$, any nonnegative proportional function $cp_X(\cdot)$ could be used. Gelman (1992) describes an iterative variation in which n is allowed to increase as m increases; that is, as the sampling continues, more variates are generated from the distribution with density g_Y.

2.9   Methods for Distributions with Certain Special Properties

Because of the analytical and implementation burden involved in building a random number generator, a general rule is that a single algorithm that works in two settings is better than two different algorithms, one for each setting. This is true, of course, unless the individual algorithms perform better in the respective special cases, and then the question is how much better. In random number generation from nonuniform distributions, it is desirable to have "universal algorithms" that use general methods that we have discussed above, yet that are optimized for certain broad classes of distributions.

For distributions with certain special properties, general algorithms using mixtures and rejection can be optimized for broad classes of distributions. We have already discussed densities that are nearly linear (Algorithm 2.5, page 51) and densities that are nearly flat (Algorithm 2.9, page 58).

Another special property of some distributions is log-concavity. A distribution is log-concave if its density (or probability function) has the property

$$\log p(x_1) - 2\log p\left(\frac{x_1 + x_2}{2}\right) + \log p(x_2) < 0,$$

whenever the densities are positive. Devroye (1986), describes general methods for a log-concave distribution, and Devroye (1987) describes a method for a discrete distribution that is log-concave. Gilks (1992) and Gilks and Wild (1992) describe a method they call *adaptive rejection sampling* or *ARS* for a continuous log-concave distribution. Many of the commonly used distributions, such as the normal, the gamma with shape parameter greater than 1, and the beta with parameters greater than 1 are log-concave. (See Dellaportas and Smith, 1993, for some examples in generalized linear models.)

The adaptive rejection method described by Gilks (1992) begins with a set S_k consisting of the points $x_0 < x_1 < \ldots < x_k < x_{k+1}$, from the range of the distribution of interest. Define L_i as the straight line determined by the points $(x_i, \log p(x_i))$ and $(x_{i+1}, \log p(x_{i+1}))$; then, for $i = 1, 2, \ldots, k$, define the piecewise linear function $h_k(x)$ as

$$h_k(x) = \min\left(L_{i-1}(x),\ L_{i+1}(x)\right), \quad \text{for } x_i \le x < x_{i+1}.$$

This piecewise linear function is a majorizing function for the log of the density, as shown in Figure 2.13.

The chords formed by the continuation of the line segments form functions that can be used as a squeeze function, $m_k(x)$, which is also piecewise linear.

For the density itself, the majorizing function and the squeeze function are piecewise exponentials. The majorizing function is

$$cg_k(x) = \exp h_k(x),$$

where each piece of $g_k(x)$ is an exponential density function truncated to the appropriate range. The density is shown in Figure 2.14.

In each step of the acceptance/rejection algorithm, the set S_k is augmented by the point generated from the majorizing distribution and k is increased by 1. The method is shown in Algorithm 2.16. In Exercise 2.13, page 82, you are asked to write a program for performing adaptive rejection sampling for the density shown in Figure 2.14, which is the same one as in Figure 2.3, and to compare the efficiency of this method with the standard acceptance/rejection method.

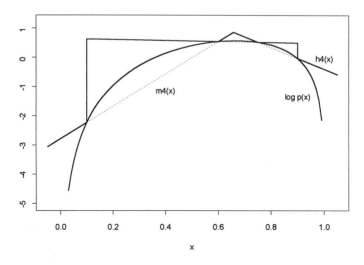

Figure 2.13: Adaptive Majorizing Function with the Log-Density (Same Density as in Figure 2.3)

Algorithm 2.16 Adaptive Acceptance/Rejection Sampling

0. Initialize k and S_k.

1. Generate y from g_k.

2. Generate u from a uniform $(0,1)$ distribution.

3. If $u \leq \dfrac{\exp m_k(y)}{cg_k(y)}$, then
   3.a. deliver y;
   otherwise
         3.b. if $u \leq \dfrac{p(y)}{cg_k(y)}$, deliver y;
         3.c. set $k = k + 1$; add y to S_k; and update h_k, g_k, and m_k.

4. Go to step 1.   ∎

After an update step the new piecewise linear majorizing function for the log of the density is as shown in Figure 2.15.

Gilks and Wild (1992) describe a similar method, but instead of using secants as the piecewise linear majorizing function, they use tangents of the log of the density. This requires computation of numerical derivatives of the log density.

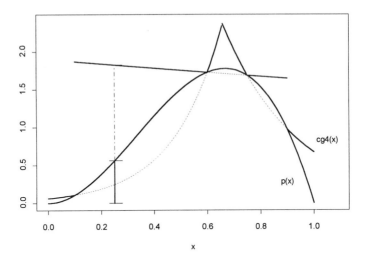

Figure 2.14: Exponential Adaptive Majorizing Function with the Density in Figure 2.3

Hörmann (1994a) adapts the methods of Gilks (1992) and Gilks and Wild (1992) to discrete distributions. Hörmann (1995) extends these methods to a distribution whose density p can be transformed by T, such that $T(p(x))$ is concave. In that case, the density p is said to be "T-concave". Often a good choice is $T = -1/\sqrt{x}$. This transformation allows construction of a table-mountain majorizing function (reminiscent of a majorizing function in the ratio-of-uniforms method) that is then used in an acceptance/rejection method. Hörmann calls this method *transformed density rejection*. Gilks, Best, and Tan (1995; corrigendum, Gilks, Neal, Best, and Tan, 1997) develop an adaptive rejection method that does not require the density to be log-concave. They call the method adaptive rejection Metropolis sampling.

Many densities of interest are unimodal. Marsaglia and Tsang (1984) give other algorithms generally applicable to unimodal distributions. The methods involve forming two regions, one on each side of the mode. A T-concave distribution is unimodal, and hence, the general methods apply to a T-concave distribution. The adaptive acceptance/rejection method is usually more efficient, however, because the number of regions grows in such a way that the majorizing function approaches the log-density.

Another broad class of distributions are those that are infinitely divisible. Damien, Laud, and Smith (1995) give general methods for generation of random deviates from distributions that are infinitely divisible.

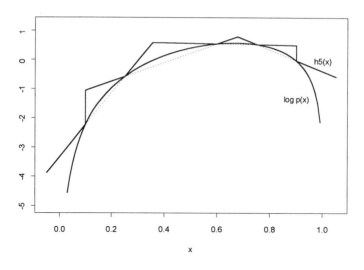

Figure 2.15: Adaptive Majorizing Function with an Additional Point

2.10 General Methods for Multivariate Distributions

The two most common methods of generating multivariate random variates make use of variates from univariate distributions. One way is to generate a vector of i.i.d. variates, and then apply a transformation to yield a vector from the desired multivariate distribution. Another way is to generate a marginal and then a sequence of conditions. This method uses the representation of the distribution function or density function as a product of the form

$$p_{X_1 X_2 X_3 \cdots X_d} = p_{X_1 | X_2 X_3 \cdots X_d} \cdot p_{X_2 | X_3 \cdots X_d} \cdot p_{X_3 | \cdots X_d} \cdots p_{X_d}.$$

We see two simple examples of these methods at the beginning of Section 3.2, page 105. In the first example in that section, we generate a d-variate normal with variance-covariance matrix Σ either by the transformation $x = T^{\mathrm{T}} z$, where T is a $d \times d$ matrix such that $T^{\mathrm{T}} T = \Sigma$ and z is a d-vector of i.i.d. $N(0, 1)$ variates. In the second case, we generate x_1 from $N_1(0, \sigma_{11})$, then generate x_2 conditionally on x_1, then generate x_3 conditionally on x_1 and x_2, and so on.

As mentioned in discussing acceptance/rejection methods in Sections 2.2 and 2.7, these methods are directly applicable to multivariate distributions, so acceptance/rejection is a third general way of generating multivariate observations. As in the example of the bivariate gamma on page 53, however, this usually involves a multivariate majorizing function, so we are still faced with the basic problem of generating from some multivariate distribution.

An iterative method, somewhat similar to the use of marginals and conditionals, can also be used to generate multivariate observations. This method was used by Geman and Geman (1984) for generating observations from a Gibbs distribution (Boltzmann distribution), and so is called the *Gibbs method.* In the Gibbs method, after choosing a starting point, the components of the d-vector variate are generated one at a time conditionally on all others. If p_X is the density of the d-variate random variable X, we use the conditional densities $p_{X_1|X_2,X_3,\cdots,X_d}$, $p_{X_2|X_1,X_3,\cdots,X_d}$, and so on. At each stage the conditional distribution uses the most recent values of all the other components. Obviously, it may require a number of iterations before the choice of the initial starting point is washed out.

The method is shown in Algorithm 2.17. (In the algorithms to follow, we represent the support of the density of interest by S, where $S \subseteq \mathbb{R}^d$.)

Algorithm 2.17 Gibbs Sampling

0. Set $k = 0$.

1. Choose $x^{(k)} \in S$.

2. Generate $x_1^{(k+1)}$ conditionally on $x_2^{(k)}, x_3^{(k)}, \ldots, x_d^{(k)}$,
   Generate $x_2^{(k+1)}$ conditionally on $x_1^{(k+1)}, x_3^{(k)}, \ldots, x_d^{(k)}$,

   . . .

   Generate $x_{d-1}^{(k+1)}$ conditionally on $x_1^{(k+1)}, x_2^{(k+1)}, \ldots, x_d^{(k)}$,
   Generate $x_d^{(k+1)}$ conditionally on $x_1^{(k+1)}, x_2^{(k+1)}, \ldots, x_{d-1}^{(k+1)}$.

3. If convergence has occurred, then
   3.a. deliver $x = x^{(k+1)}$;
   otherwise
   3.b. set $k = k + 1$, and go to step 2. ∎

Casella and George (1992) give a simple proof that this iterative method converges; that is, as $k \to \infty$, the density of the realizations approaches p_X. The question of whether convergence has practically occurred in a finite number of iterations is similar in the Gibbs method to the same question in the Metropolis-Hastings method discussed in Section 2.7. In either case, to determine that convergence has occurred is not a simple problem.

Once a realization is delivered in Algorithm 2.17, that is, once convergence has been deemed to have occurred, subsequent realizations can be generated either by starting a new iteration with $k = 0$ in step 0, or by continuing at step 1 with the current value of $x^{(k)}$. If the chain is continued at the current value of $x^{(k)}$, we must remember that the subsequent realizations are not independent. This affects variance estimates (second order sample moments), but not means (first order moments). In order to get variance estimates we may use means of batches of subsequences or use just every m^{th} (for some $m > 1$) deviate in step 3. (The idea is that this separation in the sequence will yield subsequences or a systematic subsample with correlations nearer 0. See Section 5.2 for a description of batch means.) If we just want estimates of means, however, it is best not to subsample the sequence; that is, the variances of the estimates of means (first order sample moments) using the full sequence is smaller than the variances of the estimates of the same means using a systematic (or any other) subsample (so long as the Markov chain is stationary.)

To see this, let $\bar{x}_i$ be the mean of a systematic subsample of size n consisting of every m^{th} realization beginning with the ith realization of the converged sequence. Now, following MacEachern and Berliner (1994), we observe that

$$|\text{Cov}(\bar{x}_i, \bar{x}_j)| \le \text{V}(\bar{x}_l)$$

for any positive i, j, and l less than or equal to m. Hence if $\bar{x}$ is the sample mean of a full sequence of length nm, then

$$
\begin{aligned}
\text{V}(\bar{x}) &= \text{V}(\bar{x}_l)/m + \sum_{i \ne j; i, j = 1}^{m} \text{Cov}(\bar{x}_i, \bar{x}_j)/m^2 \\
&\le \text{V}(\bar{x}_l)/m + m(m-1)\text{V}(\bar{x}_l)/m \\
&= \text{V}(\bar{x}_l).
\end{aligned}
$$

See also Geyer (1992) for discussion of subsampling in the chain.

The paper by Gelfand and Smith (1990) was very important in popularizing the Gibbs method. Gelfand and Smith also describe a related method of Tanner and Wong (1987), called *data augmentation*, which Gelfand and Smith call *substitution sampling*. In this method, a single component of the d-vector is chosen (in step 1), and then multivariate subvectors are generated conditional on just one component. This method requires $d(d-1)$ conditional distributions. The reader is referred to their article and to Schervish and Carlin (1992) for descriptions and comparisons with different methods. Tanner (1996) defines a *chained data augmentation*, which is the Gibbs method described above.

In the Gibbs method the components of the d-vector are changed systematically, one at a time. The method is sometimes called *alternating conditional sampling* to reflect this systematic traversal of the components of the vector.

Another type of Metropolis method is the hit-and-run sampler. In this method all components of the vector are updated at once. The method is shown in Algorithm 2.18, in the general version described by Chen and Schmeiser (1996).

Algorithm 2.18 Hit-and-Run Sampling

0. Set $k = 0$.

1. Choose $x^{(k)} \in S$.

2. Generate a random normalized direction $v^{(k)}$ in $\mathbb{R}^d$. (This is equivalent to a random point on a sphere, as discussed on page 109.)

3. Determine the set $S^{(k)} \subseteq \mathbb{R}$ consisting of all $\lambda \ni (x^{(k)} + \lambda v^{(k)}) \in S$. ($S^{(k)}$ is one-dimensional; S is d-dimensional.)

4. Generate $\lambda^{(k)}$ from the density $g^{(k)}$, which has support $S^{(k)}$.

5. With probability $a^{(k)}$,
   5.a. set $x^{(k+1)} = x^{(k)} + \lambda^{(k)} v^{(k)}$;
   otherwise
   5.b. set $x^{(k+1)} = x^{(k)}$.

6. If convergence has occurred, then
   6.a. deliver $x = x^{(k+1)}$;
   otherwise
   6.b. set $k = k + 1$, and go to step 2. ∎

Chen and Schmeiser (1996) discuss various choices for $g^{(k)}$ and $a^{(k)}$. One choice is

$$
g^{(k)}(\lambda) = \begin{cases} \dfrac{p(x^{(k)} + \lambda v^{(k)})}{\int_{S^{(k)}} p(x^{(k)} + u v^{(k)}) \, du} & \text{for } \lambda \in S^{(k)}, \\[2ex] 0 & \text{otherwise,} \end{cases}
$$

and

$$
a^{(k)} = 1.
$$

Another choice is g uniform over $S^{(k)}$ if $S^{(k)}$ bounded, or else some symmetric distribution centered on 0 (such as a normal or Cauchy), together with

$$
a^{(k)} = \min\left(1, \frac{p(x^{(k)} + \lambda^{(k)} v^{(k)})}{p(x^{(k)})}\right).
$$

Smith (1984) uses the hit-and-run sampler for generating uniform points over bounded regions; and Bélisle, Romeijn, and Smith (1993) use it for generating random variates from general multivariate distributions. Proofs of the convergence of the method can be found in Bélisle, Romeijn, and Smith (1993) and Chen and Schmeiser (1996).

Berbee et al. (1987) compare the efficiency of hit-and-run methods with acceptance/rejection methods and find the hit-and-run methods to be more efficient in higher dimensions. Gilks, Roberts, and George (1994) describe a generalization of the hit-and-run algorithm called *adaptive direction sampling*. In this method a set of current points is maintained, and only one, chosen at random from the set, is updated at each iteration (see Gilks and Roberts, 1996).

Both the Gibbs method and hit-and-run are special cases of the Metropolis/Hastings method in which the r of step 2 in Algorithm 2.15 (page 67) is exactly 1, so there is never a rejection.

The same issues of convergence that we encountered in discussing the Metropolis/Hastings methods must be addressed when using the Gibbs or hit-and-run methods. The need to run long chains can increase the number of computations to unacceptable levels. Schervish and Carlin (1992) and Cowles and Carlin (1996) discuss general conditions for convergence of the Gibbs sampler. Dellaportas (1995) discusses some issues in the efficiency of random number generation using the Gibbs method. See Casella and George (1992) for a description of the Gibbs sampler; and see Chen and Schmeiser (1993) for some general comparisons of Gibbs, hit-and-run, and variations. Generalizations about the performance of the methods are difficult; the best method often depends on the problem.

2.11   Generating Samples from a Given Distribution

Usually in applications, rather than just generating a single random deviate, we generate a random sample of deviates from the distribution of interest. A random sample of size n from a discrete distribution with probability function

$$p(X = m_i) = p_i$$

has a vector of counts of the mass points that has a multinomial $(n, p_1, \ldots, p_k)$ distribution.

If the sample is to be used as a set, rather than as a sequence, and if n is large relative to k, it obviously makes more sense to generate a single multinomial $(x_1, x_2, \ldots, x_k)$ and use these values as counts of occurrences of the respective mass points $m_1, m_2, \ldots, m_k$. (Methods for generating multinomials are discussed in Section 3.2.2, page 106.)

This same idea can be applied to continuous distributions with a modification to discretize the range (see Kemp and Kemp, 1987).

Exercises

2.1. The inverse CDF method.

   (a) Prove that if X is a random variable with an absolutely continuous distribution function P_X, the random variable $P_X(X)$ has a $U(0, 1)$ distribution.

   (b) Prove that the inverse CDF method for discrete random variables as specified in the relationship in expression (2.1) on page 43 is correct.

2.2. Formally prove that the random variable delivered in Algorithm 2.4 on page 48 has the density p_X. *Hint:* For the delivered variable, Z, determine the distribution function $\Pr(Z \leq x)$ and differentiate.

2.3. Write a Fortran or C function to implement the acceptance/rejection method for generating a beta(3,2) random deviate. Use the majorizing function shown in Figure 2.3 on page 48. The value of c is 1.2. Use the inverse CDF method to generate a deviate from g. (This will involve taking a square root.)

2.4. Acceptance/rejection methods.

(a) Give an algorithm to generate a normal random deviate using the acceptance/rejection method with the double exponential density (see equation (3.4), page 93) as the majorizing density. After you have obtained the acceptance/rejection test, try to simplify it.

(b) Write a program to generate bivariate normal deviates with mean $(0,0)$, variance $(1,1)$, and correlation ρ. Use a bivariate product double exponential density as the majoring density. Now set $\rho = 0.5$ and generate a sample of 1,000 bivariate normals. Compare the sample statistics with the parameters of the simulated distribution.

2.5. What would be the problem with using a normal density to make a majorizing function for the double exponential distribution (or using a half-normal for an exponential)?

2.6. (a) Write a Fortran or C function to implement the acceptance/rejection method for a bivariate gamma (2.2), page 53, using the method described in the text. (You must develop a method for determining the mode.)

(b) Now, instead of the bivariate uniform in the rectangle near the origin, devise a pyramidal distribution to use as a majorizing density.

(c) Use Monte Carlo to compare efficiency of the method using the bivariate uniform, and the method using a pyramidal density.

2.7. Consider the acceptance/rejection method given in Algorithm 2.4 to generate a realization of a random variable X with density function p_X, using a density function g_Y.

(a) Let T be the number of passes through the three steps until the desired variate is delivered. Determine the mean and variance of T (in terms of p_X and g_Y).

(b) Now consider a modification of the rejection method in which steps 1 and 2 are reversed, and the branch in step 3 is back to the new step 2, that is:

1. Generate u from a uniform (0,1) distribution.
2. Generate y from the distribution with density function g_Y.
3. If $u \leq p_X(y)/cg_Y(y)$, then take y as the desired realization; otherwise return to step 2.

Is this a better method? Let Q be the number of passes through these three steps until the desired variate is delivered. Determine the mean and variance of Q. (This method was suggested by Sibuya, 1961, and analyzed by Greenwood, 1976c.)

2.8. Formally prove that the random variable delivered in Algorithm 2.5 on page 51 has the density p.

2.9. Write a Fortran of C function to implement the ratio-of-uniforms method (page 59) to generate deviates from a gamma distribution with shape parameter α. Generate a sample of size 1,000 and perform a chi-squared goodness-of-fit test. (See Cheng and Feast, 1979.)

2.10. Use the Metropolis-Hastings algorithm (page 67) to generate a sample of standard normal random variables. Use as the candidate generating density, $g(x|y)$, a normal density in x with mean y. Experiment with different burn-in periods and different starting values. Plot the sequences generated. Test your samples for goodness-of-fit to a normal distribution. (Remember they are correlated.) Experiment with different sample sizes.

2.11. Obtain a sample of size 100 from the beta(3,2) distribution, using the SIR method of Section 2.8, using a sample of size 1,000 from the density g_Y that is proportional to the triangular majorizing function used in Exercise 2.3. (Use Algorithm 4.1, page 122, to generate the sample without replacement.) Compare the efficiency of the program you have written with the one you wrote in Exercise 2.3.

2.12. Formally prove that the random variable delivered in Algorithm 2.16 on page 74 has the density p_X. (Compare Exercise 2.2.)

2.13. Write a computer program to implement the adaptive acceptance/rejection method for generating a beta(3,2) random deviate. Use the majorizing function shown in Figure 2.13 on page 74. The initial value of k is 4, and $S_k = \{0.00, 0.10, 0.60, 0.75, 0.90, 1.00\}$. Compare the efficiency of the program you have written with the ones you wrote in Exercises 2.3 and 2.11.

2.14. Consider the trivariate normal distribution used as the example in Figure 2.12 (page 71).

   (a) Use the Gibbs method to generate and plot 1,000 realizations of X_1 (including any burn-in). Explain any choices you make in how to proceed with the method.

   (b) Use the hit-and-run method to generate and plot 1,000 realizations of X_1 (including any burn-in). Explain any choices you make in how to proceed with the method.

   (c) Compare the Metropolis-Hastings method (page 71), and the Gibbs and the hit-and-run methods for this problem.

2.15. Consider a probability model in which the random variable X has a binomial distribution with parameters n and y, which are, respectively, realizations of a conditional shifted Poisson distribution and a conditional beta distribution. For fixed λ, α, and β, let the joint density of X, N, and Y be proportional to

$$\frac{\lambda^n}{x!(n-x)!} y^{x+\alpha-1}(1-y)^{nx+\beta-1}e^{-\lambda}, \quad \text{for} \quad x = 0,1,\ldots,n;$$

$$0 \le y \le 1;$$
$$n = 1,2,\ldots.$$

First determine the conditional densities for $X|y,n$, $Y|x,n$, and $N|x,y$. Next, write a Fortran or C program to sample X from the multivariate distribution for given λ, α, and β. Now, set $\lambda = 16$, $\alpha = 2$, and $\beta = 4$, run 500 independent Gibbs sequences of length $k = 10$, taking only the final variate, plot a histogram of the observed x. (Use a random starting point.) Now repeat, except use only one Gibbs sequence of length 5,000, and plot a histogram of all observed x's after the ninth one. (See Casella and George, 1992.)

2.16. Generate a random sample of 1,000 Bernoulli variates, with $\pi = 0.3$. Do not use Algorithm 2.3; use the method of Section 2.11.

Chapter 3

Simulating Random Numbers from Specific Distributions

For the important distributions, specialized algorithms based on the general methods discussed in the previous chapter are available. The important difference in the algorithms is their speed. A secondary difference is the size and complexity of the program to implement the algorithm. Because all of the algorithms for generating from nonuniform distributions rely on programs to generate from uniform distributions, an algorithm that uses only a small number of uniforms to yield a variate of the target distribution may be faster on a computer system on which the generation of the uniform is very fast. As we have mentioned, on a given computer system there may be more than one program available to generate uniform deviates. Often a portable generator is slower than a nonportable one, so for portable generators of nonuniform distributions those that require a small number of uniform deviates may be better. If evaluation of elementary functions is a part of the algorithm for generating random deviates, then the speed of the overall algorithm depends on the speed of the evaluation of the functions. The relative speed of elementary function evaluation is different on different computer systems.

The algorithm for a given distribution is some specialized version of those discussed in the previous chapter. Often the algorithm uses some combination of these general techniques.

Many algorithms require some setup steps to compute various constants and to store tables; therefore, there are two considerations for the speed: the setup time and the generation time. In some applications many random numbers from the same distribution are required. In those cases the setup time may not be too important. In other applications the random numbers come from different distributions, probably the same family of distributions, but with changing

85

values of the parameters. In those cases the setup time may be very significant. Any computation that results in a quantity that is constant with respect to the parameters of the distribution should be performed as part of the setup computations, so as to avoid performing the computation in every pass through the main part of the algorithm.

The efficiency of an algorithm may depend on the values of the parameters of the distribution. Many of the best algorithms, therefore, switch from one method to another, depending on the values of the parameters. In some cases the speed of the algorithm is independent of the parameters of the distribution. Such an algorithm is called a *uniform time* algorithm. In many cases, the most efficient algorithm in one range of the distribution is not the most efficient in other regions. Many of the best algorithms, therefore, use mixtures of the distribution.

Sometimes it is necessary to generate random numbers from some subrange of a given distribution, such as the tail region. In some cases, there are efficient algorithms for such truncated distributions. (If there is no specialized algorithm for a truncated distribution, acceptance/rejection applied to the full distribution will always work, of course.)

Methods for generating random variates from specific distributions is an area in which there have been literally hundreds of papers, each proposing some wrinkle (not always new or significant). Because the relative efficiencies ("efficiency" here means "speed") of the individual operations in the algorithms vary from one computing system to another, and also because these individual operations can be programmed in various ways, it is very difficult to compare the relative efficiencies for the algorithms. This provides fertile ground for a proliferation of "research" papers. Two other things contribute to the large numbers of insignificant papers in this area. It is easy to look at some algorithm, modify some step, and then offer the new algorithm. Thus, the intellectual capitalization required to enter the field is small. (In business and economics, this is the same reason that so many restaurants are started; only a relatively small capitalization is required.)

Another reason for the large number of papers purporting to give new and better algorithms is the diversity of the substantive and application areas that constitute the backgrounds of the authors. Monte Carlo simulation is widely used throughout both the hard and the soft sciences. Research workers in one field often are not aware of the research published in another field.

Although, of course, it is important to seek efficient algorithms, it is also necessary to consider a problem in its proper context. In Monte Carlo simulation applications, literally millions of random numbers may be generated, but the time required to generate them is likely only to be a very small fraction of the total computing time. In fact, it is probably the case that the fraction of time required for the generation of the random numbers is somehow negatively correlated with the importance of the problem. The importance of the time required to perform some task usually depends more on its proportion of the overall time of the job, rather than on its total time.

3.1 Some Specific Univariate Distributions

In this section we consider several of the more common univariate distributions and indicate methods for simulating them. The methods discussed are generally among the better ones, at least according to some criteria, but the discussion is not exhaustive. We give the details for some simpler algorithms, but in many cases the best algorithm involves many lines of a program with several constants that optimize a majorizing function or a squeeze function or the break points of mixtures. We sometimes do not describe the best method in detail, but rather refer the interested reader to the relevant literature. Devroye (1986) has given a comprehensive treatment of methods for generating deviates from various distributions, and more information on many of the algorithms in this section can be found in that reference. In the descriptions of the algorithms, we do not identify the computations that should be removed from the main body of the algorithm and made part of some setup computations.

Two variations of a distribution are often of interest. In one variation, the distribution is truncated. In this case, the range of the original distribution is restricted to a subrange, and the probability measure adjusted accordingly. In another variation, the role of the random variable and the parameter of the distribution are interchanged. In some cases these quantities have a natural association, and the corresponding distributions are said to be conjugate. An example of two such distributions are the binomial and the beta. What is a realization of a random variable in one distribution is a parameter in the other distribution. For many distributions, we may want to generate samples of a parameter, given realizations of the random variable (the data).

3.1.1 Standard Distributions and Folded Distributions

A distribution can sometimes be simplified by transformations of the random variable that effectively remove certain parameters that characterize the distribution. In many cases the algorithms for generating random deviates address the simplified version of the distribution. An appropriate transformation is then applied to yield deviates from the distribution with the given parameters.

A linear transformation, $Y = aX + b$, is simple to apply, and is one of the most useful. The multiplier affects the *scale* and the addend affects the *location*. For example, a "three-parameter" gamma distribution, with density,

$$p(y) = \frac{1}{\Gamma(\alpha)\beta^\alpha}(y - \gamma)^{\alpha-1}e^{-(y-\gamma)/\beta}, \quad \text{for } \gamma \leq y \leq \infty,$$

can be formed from the simpler distribution with density

$$g(x) = \frac{1}{\Gamma(\alpha)}x^{\alpha-1}e^{-x}, \quad \text{for } 0 \leq x \leq \infty$$

using the transformation $Y = \beta X + \gamma$. The β parameter is a scale parameter, and γ is a location parameter. (The remaining α parameter is called the "shape parameter", and it is the essential parameter of the family of gamma distributions.) The simpler form is called the *standard* gamma distribution. Other distributions have similar standard forms.

For symmetric distributions, a useful nonlinear transformation is the absolute value. The distribution of the absolute value is often called a "folded" distribution. The exponential distribution, for example, is the folded double exponential distribution (see page 92).

3.1.2 Normal Distribution

The normal distribution, which we denote by $N(\mu, \sigma^2)$, has the probability density

$$p(x) = \frac{1}{\sqrt{2\pi}\sigma} e^{-(x-\mu)^2/(2\sigma^2)}, \quad \text{for } -\infty \leq x \leq \infty.$$

If $Z \sim N(0,1)$ and $X = \sigma Z + \mu$, then $X \sim N(\mu, \sigma^2)$. Because of this simple relationship, it is sufficient to develop methods to generate deviates from the *standard* normal distribution, $N(0,1)$, and so there is no setup involved. All constants necessary in any algorithm can be precomputed and stored.

There are several methods for transforming uniform random variates into normal random variates.

One transformation *not* to use:

if $\{U_i;\ i = 1, 12\}$ i.i.d. U(0,1), then

$$X = \sum U_i - 6$$

has an approximate normal (0,1) distribution.

This method is the Central Limit Theorem applied to a sample of size 12. Not only is the method approximate, it is also slower than better methods.

A simple and good method is the Box-Muller method arising from a polar transformation: If U_1 and U_2 are independently distributed as uniform (0,1), and

$$
\begin{aligned}
X_1 &= \sqrt{-2\log(U_1)} \cos(2\pi U_2) \\
X_2 &= \sqrt{-2\log(U_1)} \sin(2\pi U_2)
\end{aligned}
\tag{3.1}
$$

then X_1 and X_2 are independently distributed as normal (0,1) (see Exercises 3.1a and 3.1b on page 118). This method is frequently maligned by people who analyze it with inappropriate uniform generators. For example, Neave (1973) shows that when the uniform generator is a congruential generator with a very small multiplier, the resulting normals are deficient in the tails; and Golder and Settle (1976) under similar conditions demonstrate that the density of the generated normal variates has a jagged shape, especially in the tails. Of course, if they had analyzed their small-multiplier congruential generator, they would have found it lacking. (See the discussion about Figure 1.2, page 10.) Tezuka (1991) shows that similar effects also are noticeable if a poor Tausworthe generator is used.

Bratley, Fox, and Schrage (1987) show that the generated normal variates lie pairwise on spirals, and say because of this, as "an approximation to a pair of *independent* variates [the method] is terrible". The spirals, however, are of exactly the same origin as the lattice of the congruential generator itself.

To alleviate potential problems of patterns in the output of a polar method such as the Box-Muller transformation, some authors have advocated that for each pair of uniforms, only one of the resulting pair of normals be used. If there is any marginal gain in quality, it is generally not noticeable.

The Box-Muller transformation is one of several polar methods. They all have similar properties, but the Box-Muller transformation generally requires slower computations. Although most currently available computing systems can evaluate the necessary trigonometric functions extremely rapidly, the Box-Muller transformation can often be performed more efficiently using an acceptance/rejection algorithm, as we indicated in the general discussion of acceptance/rejection methods (see Exercise 3.1d on page 118). The Box-Muller transformation is implemented via rejection in Algorithm 3.1.

Algorithm 3.1 A Rejection Polar Method for Normal Variates

1. Generate v_1 and v_2 independently from $U(-1, 1)$, and set $r^2 = v_1^2 + v_2^2$.

2. If $r^2 \geq 1$, then
   go to step 1;
   otherwise
   deliver
$$x_1 = v_1 \sqrt{-2 \log r^2 / r^2}$$
$$x_2 = v_2 \sqrt{-2 \log r^2 / r^2}.$$ ∎

Ahrens and Dieter (1988) describe fast polar methods for the Cauchy and exponential distributions in addition to the normal.

The fastest algorithms for generating normal deviates use either a ratio-of-uniforms method or a mixture with acceptance/rejection. One of the best algorithms, called the rectangle/wedge/tail method, is described by Marsaglia, MacLaren, and Bray (1964). In that method the normal density is decomposed into a mixture of densities with shapes as shown in Figure 3.1. It is easy to generate a variate from one of the rectangular densities, so the decomposition is done to give a high probability of being able to use a rectangular density. That, of course, means lots of rectangles, which brings some inefficiencies. The optimal decomposition must address those trade-offs. The wedges are nearly linear densities (see Algorithm 2.5), so generating from them is relatively fast. The tail region takes the longest time, so the decomposition is such as to give a small probability to the tail. Ahrens and Dieter (1972) give an implementation of the rectangle/wedge/tail method that can be optimized at the bit level.

Kinderman and Monahan (1977), when they first introduced the ratio-of-uniforms method, applied it to the normal distribution. Leva (1992a) gives an

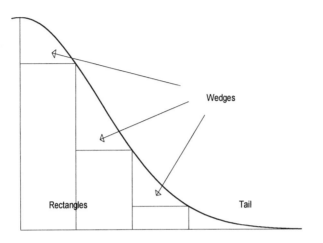

Figure 3.1: Rectangle/Wedge/Tail Decomposition

algorithm based on improved bounding curves in the ratio-of-uniforms method. (The 15-line Fortran program implementing Leva's method is Algorithm 712 of *CALGO*; see Leva, 1992b.)

Kinderman and Ramage (1976) represent the normal density as a mixture, and apply a variety of acceptance/rejection and table look-up techniques for the components. The individual techniques for various regions have been developed by Marsaglia (1964), Marsaglia and Bray (1964), and Marsaglia, MacLaren, and Bray (1964). Marsaglia and Tsang (1984) also give a decomposition, resulting in what they call the "ziggurat method".

Given the current speed of the standard methods of evaluating the inverse normal CDF, the inverse CDF method is often useful, especially if order statistics are of interest. Even with the speed of the standard algorithms for the inverse normal CDF, specialized versions, possibly to a slightly lower accuracy, have been suggested, for example by Marsaglia (1991) and Marsaglia, Zaman, and Marsaglia (1994). (The latter reference gives two algorithms for inverting the normal CDF; one very accurate, and one faster but slightly less accurate.)

Wallace (1996) describes an interesting method of generating normals from other normals, rather than by making explicit transformations of uniforms. The method begins with a set of kp normal deviates, generated by some standard method. The deviates are normalized so that their sum of squares is 1024. Let X be a $k \times p$ array containing those deviates and let A_i be a $k \times k$ orthogonal matrix. New normal deviates are formed by multiplication of an orthogonal matrix and the columns of X. A random column from the array is chosen and a random method of moving from one column to another is chosen. In Wallace's implementation, k is 4 and p is 256. Four orthogonal 4×4 matrices are chosen so as to make the matrix/vector multiplication fast:

$$A_1 = \frac{1}{2} \begin{bmatrix} 1 & 1 & -1 & 1 \\ 1 & -1 & 1 & 1 \\ 1 & -1 & -1 & -1 \\ -1 & -1 & -1 & 1 \end{bmatrix} \qquad A_2 = \frac{1}{2} \begin{bmatrix} 1 & -1 & -1 & -1 \\ 1 & -1 & 1 & 1 \\ 1 & 1 & -1 & 1 \\ -1 & -1 & -1 & 1 \end{bmatrix}$$

$$A_3 = \frac{1}{2} \begin{bmatrix} 1 & -1 & 1 & 1 \\ -1 & -1 & 1 & -1 \\ -1 & -1 & -1 & 1 \\ -1 & 1 & 1 & 1 \end{bmatrix} \qquad A_4 = \frac{1}{2} \begin{bmatrix} -1 & 1 & -1 & -1 \\ -1 & -1 & 1 & -1 \\ -1 & 1 & 1 & 1 \\ 1 & 1 & 1 & -1 \end{bmatrix},$$

hence, the matrix multiplication is usually just the addition of two elements of the vector. After a random column of X is chosen (that is, a random integer between 1 and 256), a random odd number between 1 and 255 is chosen as a stride, that is, as an increment for the column number, to allow movement from one column to another. The first column chosen is multiplied by A_1, the next by A_2, the next by A_3, the next by A_4, and then the next by A_1, and so on. The elements of the vectors that result from these multiplications constitute both the normal deviates output in this pass and the elements of a new $k \times p$ array. Except for rounding errors, the elements in the new array should have a sum of squares of 1024 also. Just to avoid any problems from rounding, however, the last element generated is not delivered as a normal deviate, but instead is used to generate a chi-squared deviate, y, with 1024 degrees of freedom via a Wilson-Hilferty approximation, and the 1023 other values are normalized by $\sqrt{y/1024}$. (The Wilson-Hilferty approximation relates the chi-squared random variable Y with ν degrees of freedom to the standard normal random variable X by

$$X \approx \frac{\left(\frac{Y}{\nu}\right)^{\frac{1}{3}} - \left(1 - \frac{2}{9\nu}\right)}{\sqrt{\frac{2}{9\nu}}}.$$

The approximation is fairly good for $\nu > 30$. See Abramowitz and Stegun, 1964.)

In Monte Carlo studies, the tail behavior is often of interest. Variates from the tail of a distribution can always be formed by selecting variates generated from the full distribution, of course; but this can be a very slow process. Marsaglia (1964), Geweke (1991a), and Robert (1995) give methods for generating directly from a truncated normal distribution. The truncated normal with left truncation point τ has density

$$p(x) = \frac{e^{-(x-\mu)^2/(2\sigma^2)}}{\sqrt{2\pi}\sigma \left(1 - \Phi\left(\frac{\tau-\mu}{\sigma}\right)\right)}, \qquad \text{for } \alpha \leq x \leq \infty,$$

where $\Phi(\cdot)$ is the standard normal CDF. The method of Robert uses an acceptance/rejection method with a translated exponential as the majorizing density, that is,

$$g(x) = \lambda^* e^{-\lambda^*(x-\tau)}, \quad \text{for } \tau \le x \le \infty,$$

where

$$\lambda^* = \frac{\tau + \sqrt{\tau^2 + 4}}{2}. \tag{3.2}$$

3.1.3 Exponential, Double Exponential, and Exponential Power Distributions

The exponential distribution with parameter $\lambda > 0$ has the probability density

$$p(x) = \lambda e^{-\lambda x}, \quad \text{for } 0 \le x \le \infty.$$

If Z has the *standard* exponential distribution, that is, with parameter equal to 1, and $X = Z/\lambda$, then X has the exponential distribution with parameter λ. Because of this simple relationship, it is sufficient to develop methods to generate deviates from the standard exponential distribution.

The inverse CDF method is very easy to implement and is generally satisfactory for the exponential distribution. The method is to generate u from $U(0,1)$ and then take

$$x = -\frac{\log(u)}{\lambda}. \tag{3.3}$$

(This and similar computations are why we require that the simulated uniform not include its endpoints.)

Many other algorithms for generating exponential random numbers have been proposed over the years. Marsaglia, MacLaren, and Bray (1964) apply the rectangle/wedge/tail method to the exponential distribution. Ahrens and Dieter (1972) give a method that can be highly optimized at the bit level.

Ahrens and Dieter also provide a catalog of other methods for generating exponentials. These other algorithms seek greater speed by avoiding the computation of the logarithm. Many simple algorithms for random number generation involve evaluation of elementary functions. As we have indicated, evaluation of an elementary function at a random point can often be performed equivalently by acceptance/rejection, and Ahrens and Dieter (1988) describe a method for the exponential that does that. As the software for evaluating elementary functions has become faster, the need to avoid their evaluation has become less.

A common use of the exponential distribution is as the model of the interarrival times in a Poisson process. A (homogeneous) Poisson process,

$$T_1 < T_2 < \dots,$$

with rate parameter λ can be generated by taking the output of an exponential random number generator with parameter λ as the times,

$$t_1, \ t_2 - t_1, \ \dots.$$

We consider nonhomogeneous Poisson processes in Section 4.3.2, page 126.

The double exponential distribution, also called the Laplace distribution, with parameter $\lambda > 0$ has the probability density

$$p(x) = \frac{\lambda}{2} e^{-\lambda|x|}, \quad \text{for} \ -\infty \le x \le \infty. \tag{3.4}$$

If Z has the standard exponential distribution and $X = SZ/\lambda$, where S is a random variable with probability mass $\frac{1}{2}$ at -1 and at $+1$, then X has the double exponential distribution with parameter λ. The double exponential distribution is often used in Monte Carlo studies of robust procedures, because it has a heavier tail than the normal distribution, yet corresponds well with observed distributions.

A generalization of the double exponential distribution is the exponential power distribution, whose density is

$$p(x) \propto e^{-\lambda|x|^\alpha}, \quad \text{for} \ -\infty \le x \le \infty.$$

The members of the family of exponential power distributions for $1 \le \alpha < 2$ are often used to model distributions with slightly heavier tails than the normal distribution. Either the double exponential or the normal distribution, depending on the value of α, works well as a majorizing density to generate exponential power variates by acceptance/rejection (see Tadikamalla, 1980a).

3.1.4 Gamma Distribution

The gamma distribution with parameters $\alpha > 0$ and β has the probability density

$$p(x) = \frac{1}{\Gamma(\alpha)\beta^\alpha} x^{\alpha-1} e^{-x/\beta}, \quad \text{for} \ 0 \le x \le \infty,$$

where $\Gamma(\alpha)$ is the complete gamma function. The α parameter is called the shape parameter, and β is called the scale parameter. If the random variable Z has the *standard* gamma distribution with shape parameter α and scale parameter 1, and $X = \beta Z$, then X has gamma distribution with parameters α and β. (Notice that the exponential is a gamma with $\alpha = 1$, and $\beta = 1/\lambda$.)

Of the special distributions we have considered so far, this is the first one that has a parameter that cannot be handled by simple translations and scalings. Hence, the best algorithms for the gamma may be different depending on the value of α and on how many deviates are to be generated for a given value of α.

Cheng and Feast (1979) and Kinderman and Monahan (1980) use the ratio-of-uniforms method for a gamma distribution with $\alpha > 1$. The method of Cheng and Feast is shown in Algorithm 3.2. The mean time of this algorithm is $O(\alpha^{\frac{1}{2}})$, so for larger values of α it is less efficient. Cheng and Feast (1980) give an acceptance/rejection method that was better for large values of the shape parameter. Schmeiser and Lal (1980) use a composition of 10 densities, some of the rectangle/wedge/tail type, followed by the acceptance/rejection method.

The Schmeiser/Lal method is the algorithm used in the IMSL Libraries for values of the shape parameter greater than 1. Sarkar (1996) gives a modification of the Schmeiser/Lal method that has greater efficiency because of using more intervals, resulting in tighter majorizing and squeeze functions, and because of using an alias method to help speed the process.

Algorithm 3.2 The Cheng/Feast (1979) Algorithm for Generating Gamma Random Variates when the Shape Parameter is Greater than 1

1. Generate u_1 and u_2 independently from $U(0,1)$, and set

$$v = \frac{\left(\alpha - \frac{1}{6\alpha}\right) u_1}{(\alpha - 1) u_2}.$$

2. If

$$\frac{2(u_2 - 1)}{\alpha - 1} + v + \frac{1}{v} \leq 2,$$

   then deliver $x = (\alpha - 1)v$;
   otherwise
      if

$$\frac{2 \log u_2}{\alpha - 1} - \log v + v \leq 1$$

   then deliver $x = (\alpha - 1)v$.

3. Go to step 1.   ∎

An efficient algorithm for values of the shape parameter less than 1 is the acceptance/rejection method described in Ahrens and Dieter (1974) and modified by Best (1983), as shown in Algorithm 3.3. That method is the algorithm used in the IMSL Libraries for values of the shape parameter less than 1.

Algorithm 3.3 The Best/Ahrens/Dieter Algorithm for Generating Gamma Random Variates when the Shape Parameter is Less than 1

0. Set $t = 0.07 + 0.75\sqrt{1 - \alpha}$ and
   $$b = \alpha + \frac{e^{-t}\alpha}{t}.$$

1. Generate u_1 and u_2 independently from $U(0,1)$, and set $v = bu_1$.

2. If $v \leq 1$, then
      set $x = tv^{\frac{1}{\alpha}}$
      if $u_2 \leq \dfrac{2 - x}{2 + x}$ then deliver x;
      otherwise
         if $u_2 \leq e^{-x}$ deliver x;
   otherwise

$$\text{set } x = \log\left(\frac{t(b-v)}{\alpha}\right) \text{ and } y = \frac{x}{t};$$

if $u_2(\alpha + y(1-\alpha)) \leq 1$ then deliver x;

otherwise

      if $u_2 \leq y^{a-1}$ deliver x.

3. Go to step 1.   ■

There are two special cases of the gamma distribution that are often used. In simulation studies, the shape parameter α often is a positive integer. In that case, the distribution is sometimes called the *Erlang distribution*. If $Y_1, Y_2, \ldots, Y_\alpha$ are independently distributed as exponentials with parameter $1/\beta$, the $X = \sum Y_i$ has a gamma (Erlang) distribution with parameters α and β. Using the inverse CDF method (equation (3.3)) with the independent realizations $u_1, u_2, \ldots, u_\alpha$, we generate an Erlang as

$$x = -\beta \log\left(\prod_{i=1}^{\alpha} u_i\right).$$

The general algorithms for gammas work better if α is large.

The other special case of the gamma is the chi-squared distribution in which the scale parameter β is 2. Twice the shape parameter α is called the degrees of freedom. For large or nonintegral degrees of freedom, the general methods for generating gamma random deviates are best for generating chi-squared deviates; otherwise, special methods described below are used.

Philippe (1997) describes a method to generate variates from a truncated gamma distribution. Many common applications require truncation on the right. The right truncated gamma has density

$$p(x) = \frac{1}{\Gamma_{\tau/\beta}(\alpha)\beta^\alpha} x^{\alpha-1} e^{-x/\beta}, \quad \text{for } 0 \leq x \leq \tau,$$

where $\Gamma_{\tau/\beta}(\alpha)$ is the incomplete gamma function. Philippe shows that if the random variable X has this distribution, it can be represented as an infinite mixture of beta random variables:

$$X = \sum_{k=1}^{\infty} \frac{\Gamma(\alpha)}{\Gamma(\alpha+k)\Gamma_{1/\beta}(\alpha)\beta^{\alpha+k-1}e^{1/\beta}} Y_k,$$

where Y_k is a random variable with a beta random variable with parameters α and k. Philippe suggested as a majorizing density a finite series

$$g_m(y) \propto \sum_{k=1}^{m} \frac{1}{\beta^{k-1}\Gamma(\alpha)\Gamma(k)\sum_{i=1}^{m}\frac{1}{\beta^{i-1}\Gamma(\alpha+i)}} x^{\alpha-1}(1-x)^{k-1}$$

$$= \sum_{k=1}^{m} \frac{1}{\beta^{k-1}\Gamma(\alpha+k)\sum_{i=1}^{m}\frac{1}{\beta^{i-1}\Gamma(\alpha+i)}} h_k(x),$$

where h_k is a beta density with parameters α and k. Thus, to generate a variate from a distribution with density g_m, we select a beta with the probability equal to the weight, and then use a method described in the next section for generating a beta variate. The number of terms depends on the probability of acceptance. Obviously, we want a high probability of acceptance, but this requires a large number of terms in the series. For a probability of acceptance of at least p^* (< 1), Philippe shows that the number of terms required in the series is approximately

$$m^* = \frac{1}{4}\left(z_p + \sqrt{z_p^2 + \frac{4}{\beta}}\right)^2,$$

where $z_p = \Phi^{-1}(p)$ and Φ is the standard normal CDF.

Algorithm 3.4 The Philippe (1997) Algorithm for Generating Gamma Random Variates Truncated on the Right at τ

0. Determine m^* and initialize quantities in g_{m^*}.

1. Generate y from the distribution with density g_{m^*}.

2. Generate u from $U(0,1)$.

3. If

$$u \leq \frac{\sum_{k=1}^{m^*}\frac{1}{\beta^{k-1}\Gamma(k)}}{e^{\frac{y}{\beta}}\sum_{k=1}^{m^*}\frac{(1-y)^{k-1}}{\beta^{k-1}\Gamma(k)}}$$

   then
       take y as the desired realization;
   otherwise
       return to step 1.   ∎

Philippe (1997) also describes methods for a left truncated gamma distribution, including special algorithms for the case when the truncation point is an integer. The interested reader is referred to the paper for the details.

3.1.5 Beta Distribution

The beta distribution with parameters $\alpha > 0$ and $\beta > 0$ has the probability density

$$p(x) = \frac{1}{B(\alpha,\beta)}x^{\alpha-1}(1-x)^{\beta-1}, \quad \text{for } 0 \leq x \leq 1, \tag{3.5}$$

where $B(\alpha,\beta)$ is the complete beta function.

Efficient methods for the beta require different algorithms for different values of the parameters. If either parameter is equal to 1, it is very simple to generate beta variates using the inverse CDF method, which in this case would just be a root of a uniform. If both values of the parameters are less than 1, the simple acceptance/rejection method of Jöhnk (1964) (Algorithm 3.5) is one of the best. If one parameter is less than 1 and the other is greater than 1, the method of Atkinson (1979) is useful. If both parameters are greater than 1, the method of Schmeiser and Babu (1980) is very efficient, except that it requires a lot of setup time. For the case of both parameters greater than 1, Cheng (1978) gives an algorithm that requires very little setup time. The IMSL Library uses all five of these methods, depending on the values of the parameters and how many deviates are to be generated for a given setting of the parameters.

Algorithm 3.5 Jöhnk's Algorithm for Generating Beta Random Variates when Both Parameters are Less than 1

1. Generate u_1 and u_2 independently from $U(0, 1)$, and set $v_1 = u_1^{1/\alpha}$ and $v_2 = u_2^{1/\beta}$.

2. Set $w = v_1 + v_2$.

3. If $w > 1$, then go to step 1.

4. Set $x = \dfrac{v_1}{w}$ and deliver x. ∎

3.1.6 Student's t, Chi-Squared, and F Distributions

The Student's t, chi-squared, and F distributions all are derived from the normal distribution. Variates from these distributions could, of course, be generated by transforming normal deviates. In the case of the chi-squared distribution, however, this would require generating and squaring several normals for each chi-squared deviate. A more direct algorithm is much more efficient. Even in the case of the t and F distributions, which would require only a couple of normals or chi-squared deviates, there are better algorithms.

Student's t Distribution

The standard t distribution with ν degrees of freedom has density

$$p(x) = \frac{\Gamma\left(\frac{\nu+1}{2}\right)}{\Gamma\left(\frac{\nu}{2}\right)\sqrt{\nu\pi}\sigma}\left(1 + \frac{x^2}{\nu}\right)^{-\frac{\nu+1}{2}} \qquad \text{for } -\infty \leq x \leq \infty.$$

The degrees of freedom, ν, does not have to be an integer, but it must be positive.

Kinderman and Monahan (1980) describe a ratio-of-uniforms method for the t distribution. The algorithm is rather complicated, but it is very efficient. Marsaglia (1980) gives a simpler procedure that is almost as fast. Either is almost twice as fast as generating a normal and dividing by the square root of a chi-squared. Marsaglia (1984) also gives a very fast algorithm for generating t variates that is based on a transformed acceptance/rejection method he called exact-approximation (see Section 2.2).

Bailey (1994) gives a polar method for the Student's t distribution that is similar to the polar method for the normal given in Algorithm 3.1.

Algorithm 3.6 A Rejection Polar Method for t Variates with ν Degrees of Freedom

1. Generate v_1 and v_2 independently from $U(-1,1)$, and set $r^2 = v_1^2 + v_2^2$.

2. If $r^2 \geq 1$, then
   go to step 1;
   otherwise

   deliver $x = v_1 \sqrt{\dfrac{\nu(r^{-4/\nu} - 1)}{r}}$. ∎

Because a t random variable with ν degrees of freedom is the square root of an F random variable with 1 and ν degrees of freedom, the relation to a beta random variable mentioned below could be used to generate t deviates.

In Bayesian analysis it is sometimes necessary to generate random variates for the degrees of freedom in a t distribution conditional on the data. In the hierarchical model underlying the analysis, the t random variable is interpreted as a mixture of normal random variables divided by square roots of gamma random variables. For given realizations of gammas, $\lambda_1, \lambda_2, \ldots, \lambda_n$, the density of the degrees of freedom is

$$p(x) \propto \prod_{i=1}^{n} \frac{\nu^{\nu/2}}{2^{\nu/2}\Gamma\left(\frac{\nu}{2}\right)} \lambda_i^{\nu/2} e^{-\nu\lambda_i/2}.$$

Mendoza-Blanco and Tu (1997) show that three different gamma distributions can be used to approximate this density very well for three different ranges of values of $\lambda_g e^{-\lambda_a}$, where λ_g is the geometric mean of the λ_i, and λ_a is the arithmetic mean. Although the approximations are very good, the gamma approximations could also be used as majorizing densities.

Chi-Squared Distribution

The chi-squared distribution, as we have mentioned above, is a special case of the gamma distribution in which the scale parameter, β, is 2. Twice the shape parameter, 2α, is called the degrees of freedom, and is often denoted by ν. If ν is large or if it is not an integer, the general methods for generating gamma random deviates described above are best for generating chi-squared deviates. If the degrees of freedom value is a small integer, the chi-squared deviates can be generated by taking a logarithm of the product of some independent uniforms. If ν is an even integer, the chi-squared deviate r is produced from $\nu/2$ independent uniforms, u_i, by

$$r = -2\log\left(\prod_{i=1}^{\nu/2} u_i\right).$$

If ν is an odd integer, this method can be used with the product going to $(\nu - 1)/2$ and then the square of an independent normal deviate is added to produce r.

The square root of the chi-squared random variable is sometimes called a chi random variable. Although, clearly, a chi random variable could be generated as the square root of a chi-squared generated as above, there are more efficient direct ways of generating a chi deviate; see Monahan (1987).

F Distribution

A variate from the F distribution can be generated as the ratio of two chi-squared deviates, which, of course, would be only half as fast as generating a chi-squared deviate. A better way to generate an F variate is as a transformed beta. If X is distributed as a beta, with parameters $\nu_1/2$ and $\nu_2/2$, and

$$Y = \frac{\nu_2}{\nu_1} \frac{X}{1 - X},$$

then Y has an F distribution with ν_1 and ν_2 degrees of freedom. Generating a beta and transforming it usually takes only slightly longer than generating a chi-squared.

3.1.7 Weibull Distribution

The Weibull distribution with parameters $\alpha > 0$ and β has the probability density

$$p(x) = \frac{\alpha}{\beta} x^{\alpha-1} e^{-x^\alpha/\beta}, \quad \text{for } 0 \leq x \leq \infty.$$

The simple inverse CDF method applied to the standard Weibull (i.e., $\beta = 1$) is quite efficient. The expression is simply

$$(-\log u)^{\frac{1}{\alpha}}.$$

Of course, an acceptance/rejection method could be used to replace the evaluation of the logarithm in the inverse CDF. The standard Weibull deviates are then scaled by β.

3.1.8 Binomial Distribution

The probability function for the binomial distribution with parameters n and π is

$$p(x) = \frac{n!}{x!(n-x)!} \pi^x (1 - \pi)^{n-x}, \quad \text{for } x = 0, 1, \ldots, n,$$

where n is a positive integer and $0 < \pi < 1$.

To generate a binomial, a simple way is to sum Bernoullis, which is equivalent to an inverse CDF technique. If n, the number of independent Bernoullis, is small, this method is adequate. The time required for this kind of algorithm is obviously $O(n)$. For larger values of n, the median of a random sample of size n from a Bernoulli distribution can be generated (it has an approximate beta distribution; see Relles, 1972), and then the inverse CDF method can be applied from that point. Starting at the median allows the time required to be halved. Kemp (1986) shows that starting at the mode results in an even faster method and gives a method to approximate the modal probability quickly. If this idea is applied recursively, the time becomes $O(\log n)$. The time required for any method based on the CDF of the binomial is an increasing function of n.

Several methods whose efficiency is not so dependent on n are available, and for large values of n they are to be preferred to methods based on the CDF. (The value of π also affects the speed; the inverse CDF methods are generally competitive so long as $n\pi < 500$.) Stadlober (1991) described an algorithm based on a ratio-of-uniforms method. Kachitvichyanukul (1982) gives an efficient method using acceptance/rejection over a composition of four regions (see Schmeiser, 1983; and Kachitvichyanukul and Schmeiser, 1988, 1990). This is the method used in the IMSL Libraries.

The beta-binomial distribution is the mixture distribution that is a binomial whose parameter π is a realization of a random variable that has a beta distribution. A simple way of generating deviates from a beta-binomial distribution is first to generate the parameter π as the appropriate beta and then to generate the binomial. In this case, an inverse CDF method may be more efficient, because it does not require as much setup time as the efficient ratio-of-uniforms of acceptance/rejection methods referred to above.

3.1.9   Poisson Distribution

The probability function for the Poisson distribution with parameter $\theta > 0$ is

$$p(x) = \frac{e^{-\theta}\theta^x}{x!}, \quad \text{for } x = 0, 1, 2, \ldots$$

A Poisson with a small mean, θ, can be generated efficiently by the inverse CDF technique.

Many of the other methods that have been suggested for the Poisson require longer times for distributions with larger means. Ahrens and Dieter (1980) and Schmeiser and Kachitvichyanukul give efficient methods whose times do not depend on the mean (see Schmeiser, 1983). The method of Schmeiser and Kachitvichyanukul uses acceptance/rejection over a composition of four regions. This is the method used in the IMSL Libraries.

3.1.10 Negative Binomial and Geometric Distributions

The probability function for the negative binomial is

$$p(x) = \binom{x + r - 1}{r - 1} (1 - \pi)^r \pi^x, \quad \text{for } x = 0, 1, 2, , \ldots,$$

where $r > 0$ and $0 < \pi < 1$. If r is an integer, the negative binomial distribution is sometimes called the Pascal distribution.

If $r\pi/(1 - \pi)$ is relatively small and $(1 - \pi)^r$ is not too small, the inverse CDF method works well. Otherwise a gamma $(r, \pi/(1 - \pi))$ can be generated and used as the parameter to generate a Poisson. The Poisson variate is then delivered as the negative binomial variate.

The geometric distribution is a special case of the negative binomial, with $r = 1$. The probability function is

$$p(x) = \pi(1 - \pi)^x, \quad \text{for } x = 0, 1, 2, \ldots$$

The simplest and one of the best methods for the geometric distribution with parameter π is to generate a uniform deviate u, and take

$$\lceil \frac{\log u}{\log(1 - \pi)} \rceil.$$

It is common to see the negative binomial and the geometric distributions defined as starting at 1 instead of 0, as above. The distributions are the same after making an adjustment of subtracting 1.

3.1.11 Hypergeometric Distribution

The probability function for the hypergeometric is

$$p(x) = \frac{\binom{M}{x} \binom{L - M}{N - x}}{\binom{L}{N}},$$

$$\text{for } x = \max(0, N - L + M), 1, 2, \ldots, \min(N, M).$$

A good method for generating from the hypergeometric distribution is the inverse CDF method. The inverse CDF can be evaluated recursively, using the simple expression for the ratio $p(x + 1)/p(x)$. Another simple method that is good is straightforward use of the finite sampling model that defines the distribution.

Kachitvichyanukul and Schmeiser (1985) give an algorithm based on acceptance/rejection of a probability function decomposed as a mixture, and Stadlober (1990) describes an algorithm based on a ratio-of-uniforms method. Both of these can be faster than the inverse CDF for larger values of N and M.

3.1.12   Logarithmic Distribution

The probability function for the logarithmic distribution with parameter θ is

$$p(x) = -\frac{a^x}{x \log(1-a)}, \quad \text{for } x = 0, 1, 2,, \ldots,$$

where $0 < \theta < 1$.

Kemp (1981) describes a method for generating efficiently from the inverse logarithmic CDF, either using a "chop down" approach to move rapidly down the set of CDF values, or using a mixture in which highly likely values are given priority.

3.1.13   Other Specific Univariate Distributions

Many other interesting distributions have simple relationships to the standard distributions discussed above. For the lognormal distribution, for example, a method that is almost as good as any special method is just to generate a normal variate and exponentiate.

Variates from the Cauchy distribution, which has density

$$p(x) = \frac{1}{\pi a \left(1 + \left(\frac{x-b}{a}\right)^2\right)}, \quad \text{for } -\infty \leq x \leq \infty,$$

can be generated easily by the inverse CDF method. For the standard Cauchy, that is, with $a = 1$ and $b = 0$, given u from $U(0,1)$, a Cauchy is $\tan(\pi u)$. The tangent function in the inverse CDF could be evaluated by acceptance/rejection in the manner discussed earlier. Kronmal and Peterson (1981) express the Cauchy distribution as a mixture, and give an acceptance/complement method that is very fast.

Other distributions for which the inverse CDF method is convenient are the Pareto distribution and the Rayleigh distribution. For the Pareto distribution, with density

$$p(x) = \frac{ab^a}{x^{a+1}}, \quad \text{for } x \geq b,$$

variates can be generated as

$$x = \frac{b}{u^{1/a}}.$$

For the Rayleigh distribution, with density

$$p(x) = \frac{x}{\sigma} e^{-x^2/2\sigma^2} \quad \text{for } x \geq 0$$

(which is a standard Weibull with shape parameter $\alpha = 2$ and scale parameter $\beta = \sqrt{2}\sigma$), variates can be generated as

$$x = \sigma \sqrt{-\log u}.$$

In both cases, faster acceptance/rejection methods can be constructed; but if the computing system has fast functions for exponentiation and taking logarithms, the inverse CDF is adequate.

There are many variations of the continuous Pareto distribution, the simplest of which is the one defined above. In addition, there are some discrete versions, including various zeta and Zipf distributions. (See Arnold, 1983, for an extensive discussion of the variations.) Variates from these distributions can usually be generated by discretizing some form of a Pareto distribution. Dagpunar (1988) describes such a method for a zeta distribution, in which the Pareto variates are first generated by an acceptance/rejection method.

Variates from the simple Zipf distribution, in which the probabilities of the integers i are proportional to i^{-a}, for $a > 1$, can be generated efficiently by a direct acceptance/rejection method given by Devroye (1986). In this method, first two variates u_1 and u_2 are generated from $U(0,1)$ and then x and t are defined as

$$x = \lfloor u_1^{-1/(a-1)} \rfloor$$

and

$$t = (1 + 1/x)^{a-1}.$$

The variate x is accepted if

$$x \leq \frac{t}{t-1} \frac{2^{a-1} - 1}{2^{a-1}} u_2.$$

Variates from the von Mises distribution, with density,

$$p(x) = \frac{1}{2\pi I_0(c)} e^{c \cos(x)}, \quad \text{for } -\pi \leq x \leq \pi,$$

as discussed on page 64, can be generated by the acceptance/rejection method. Best and Fisher (1979) use a transformed folded Cauchy distribution as the majorizing distribution. The majorizing density is

$$g(y) = \frac{1 - \rho^2}{\pi(1 + \rho^2 - 2\rho \cos y)}, \quad \text{for } 0 \leq y \leq \pi,$$

where ρ is chosen in $[0,1)$ so as to optimize the probability of acceptance for a given value of the von Mises parameter, c. This is the method used in the IMSL Libraries. Dagpunar (1990) gives an acceptance/rejection method for the von Mises distribution that is often more efficient.

A class of distributions, called D-distributions, that arise in extended gamma processes is studied by Laud, Ramgopal, and Smith (1993). The interested reader is referred to that paper for the details.

3.1.14   General Families of Univariate Distributions

General families of distributions are often useful in simulation to define distributions that seem to model observed data well. Some commonly used ones are the Pearson family, the Johnson family, the generalized lambda family, and the Burr family. The Johnson family of distributions is particularly useful because the distributions are constructed from the percentiles of the observations to be modeled. Devroye (1986) describes a method for simulating Johnson variates.

Ramberg and Schmeiser (1974) describe a generalized lambda family of distributions. Their system, which is a generalization of a system introduced by John Tukey, has four parameters that can be chosen to fit a variety of distributional shapes. They specify the distribution in terms of the inverse of its distribution function,

$$P^{-1}(u) = \lambda_1 + \frac{u^{\lambda_3} - (1 - u)^{\lambda_4}}{\lambda_2}.$$

The distribution function itself cannot be written in closed form, but the inverse allows deviates from this distribution to be generated easily by the inverse CDF method.

The Burr family of distributions is very flexible and can have a wide range of shapes (Burr, 1942). One of the common forms (Burr and Cislak, 1968) has the CDF,

$$P(x) = 1 - \frac{1}{(1 + x^\alpha)^\beta}, \quad \text{for } 0 \le x \le \infty; \ \alpha, \beta > 0, \qquad (3.6)$$

which is easily inverted.

Tadikamalla and Johnson (1982) describe a flexible general family of distributions based on transformations of logistic variables.

If the first four moments of a random variable are known, Fleishman (1978) suggested representing the random variable as a third-degree polynomial in a standard normal random variable, solving for the coefficients, and then using the resulting transformation of a normal to generate observations on the nonnormal random variable.

Tadikamalla (1980b) compares the use of these families of distributions for simulating distributions with various given shapes or given moments. As might be expected, some systems are better in one situation and others are better in other cases. The use of the polynomial of a normal random deviate, as Fleishman suggested, seems to fit a wide range of distributions, and its simplicity recommends it in many cases.

The motivation for some of the early work with general families of distributions was to use them as approximations to some standard distribution, such as a gamma, for which it is more difficult to generate deviates. As methods for the standard distributions have improved, it is more common just to generate directly from the distribution of interest. The general families, however, often provide more flexibility in choosing a distribution that matches sample data better.

Random number generation is important in studying the performance of various statistical methods, especially when the assumptions underlying the methods are not satisfied. The question is whether the statistical method is *robust*. One concern in robust statistics is how well the method would hold up in very extreme conditions, such as in the presence of a heavy-tailed distribution. The Cauchy distribution has very heavy tails; none of its moments exist. It is often used in robustness studies.

The Pareto distribution has relatively heavy tails; for some values of the parameter, the mean exists but the variance does not. A "Pareto-type" distribution is one whose distribution function satisfies the relationship

$$P(x) = 1 - x^{-\gamma} g(x),$$

where $g(x)$ is a *slowly varying function*, that is, for fixed $t > 0$,

$$\lim_{x \to \infty} \frac{g(tx)}{g(x)} = 1.$$

The Burr distribution with the CDF given in (3.6) is of the Pareto type, with $\gamma = \alpha\beta$.

The stable family of distributions is a flexible family of generally heavy-tailed distributions. This family includes the normal distribution at one extreme value of one of the parameters and the Cauchy at the other extreme value. Chambers, Mallows, and Stuck (1976) give a method for generating deviates from stable distributions. (Watch for some errors in the constants in the auxiliary function D2, for evaluating $(e^x - 1)/x$.) Their method is used in the IMSL Libraries. For a symmetric stable distribution, Devroye (1986) points out that a faster method can be developed by exploiting the relationship of the symmetric stable to the Fejer-de la Vallee Poussin distribution. Buckle (1995) shows how to simulate the parameters of a stable distribution, conditional on the data.

3.2 Some Specific Multivariate Distributions

Multivariate distributions can be built from univariate distributions either by a direct transformation of a vector of i.i.d. scalar variates or by a sequence of conditional scalar variates.

The use of various Markov chain methods in Monte Carlo has made the conditional approaches more popular in generating multivariate deviates. The hit-and-run sampler (see page 78) is particularly useful in generating from multivariate distributions (see Bélisle, Romeijn, and Smith, 1993, for example). A survey of methods and applications of multivariate simulation can be found in Johnson (1987).

3.2.1 Multivariate Normal Distribution

The d-variate normal distribution with mean vector μ and nonsingular variance-covariance matrix Σ, which we denote by $N_d(\mu, \Sigma)$, has the probability density function

$$p(x) = \frac{1}{(2\pi)^{\frac{d}{2}}|\Sigma|^{\frac{1}{2}}} \exp\left(-\frac{(x-\mu)^{\mathrm{T}}\Sigma^{-1}(x-\mu)}{2}\right).$$

A direct way of generating random vectors from the multivariate normal distribution is to generate a d-vector of i.i.d. standard normal deviates $z = (z_1, z_2, \ldots, z_d)$ and then to form the vector

$$x = T^{\mathrm{T}}z + \mu,$$

where T is a $d \times d$ matrix such that $T^{\mathrm{T}}T = \Sigma$. (T could be a Cholesky factor of Σ, for example.) Then x has a $N_d(\mu, \Sigma)$ distribution.

Another approach for generating the d-vector x from $N_d(\mu, \Sigma)$ is to generate x_1 from $N_1(0, \sigma_{11})$, generate x_2 conditionally on x_1, generate x_3 conditionally on x_1 and x_2, and so on.

Deák (1990) discribes a method for for generating multivariate normals by using a transformation to spherical coordinates together with a random orthogonal transformation. This method is also useful in evaluating multivariate normal probabilities.

To generate from a truncated multivariate normal or from a multivariate normal with other restrictions, the methods for truncated univariate normals mentioned above can be applied.

3.2.2   Multinomial Distribution

The probability function for the d-variate multinomial distribution is

$$p(x) = \frac{n!}{\prod x_j!} \prod \pi_j^{x_j}, \quad \text{for } x_j \geq 0, \text{ and } \Sigma x_j = n.$$

The parameters π_j must be positive and sum to 1.

To generate a multinomial, a simple way is to work with the marginals; they are binomials. The generation is done sequentially. Each succeeding conditional marginal is binomial. For efficiency, the first marginal considered would be the one with the largest probability.

Another interesting algorithm for the multinomial, due to Brown and Bromberg (1984), is based on the fact that the conditional distribution of independent Poisson random variables, given their sum, is multinomial. The use of this relationship requires construction of extensive tables. Davis (1993) found the Brown/Bromberg method to be slightly faster than the sequential conditional marginal binomial method, once the setup operations are performed. If multinomials are to be generated from distributions with different parameters, however, the sequential conditional marginal method is more efficient.

3.2.3 Correlation Matrices and Variance-Covariance Matrices

Generation of random correlation matrices or random variance-covariance matrices in general requires some definition of a probability measure (as generation of any random object does, of course). A probability measure for a variance-covariance matrix, for example, may be taken to correspond to that of sample variance-covariance matrices from a random sample of a given size from d-variate normal distribution with variance-covariance matrix Σ. In this case, the probability distribution, aside from a constant, is a Wishart distribution.

Hartley and Harris (1963) and Odell and Feiveson (1966) give the method in Algorithm 3.7 to generate random variance-covariance matrices corresponding to a sample of size n from a d-variate normal distribution.

Algorithm 3.7 Sample Variance-Covariance Matrices from the Multivariate Normal with Variance-Covariance Matrix Σ

0. Determine T such that $T^{\mathrm{T}}T = \Sigma$. T could be a Cholesky factor of Σ, for example.

1. Generate a sequence of i.i.d. standard normal deviates z_{ij}, for $i = 1, 2, \ldots, j$ and $j = 2, 3, \ldots, d$.

2. Generate a sequence of i.i.d. chi-squared variates, y_i, with $n - 1$ degrees of freedom.

3. Compute $B = (b_{ij})$:
$$b_{11} = y_1$$
$$b_{jj} = y_j + \sum_{i=1}^{j-1} z_{ij}^2 \quad \text{for } j = 2, 3, \ldots, d$$
$$b_{1j} = z_{1j}\sqrt{y_1}$$
$$b_{ij} = z_{ij}\sqrt{y_i} + \sum_{k=1}^{i-1} z_{ki}z_{kj} \quad \text{for } i < j = 2, 3, \ldots, d$$
$$b_{ij} = b_{ji} \quad \text{for } j < i = 2, 3, \ldots, d$$

4. Deliver $V = \frac{1}{n}T^{\mathrm{T}}BT$. ∎

The advantage of this algorithm is that it does not require n d-variate normals for each random variance-covariance matrix. The B in Algorithm 3.7 is a Wishart matrix. Smith and Hocking (1972) give a Fortran program for generating Wishart matrices using this technique. The program is available from statlib as the *Applied Statistics* algorithm AS 53. (See the bibliography.)

A random matrix from a noncentral Wishart distribution with noncentrality matrix having columns c_i can be generated from a random Wishart matrix B, using the relationship

$$S = T^{\mathrm{T}}\left(\sum_{i=1}^{d}(z_i + c_i)(z_i + c_i)^{\mathrm{T}} + B\right)T,$$

where z_i are i.i.d. random vectors from a $N_d(0, I_d)$ distribution and the z_i are independent of B. This requires d^2 additional scalar normal variates. Gleser (1976) gives a modification of this method that requires only r^2 additional normal deviates, where r is the number of linearly independent vectors c_i.

For generating random correlation matrices, again, we need some probability measure or at least some description of the space from which the matrices are to be drawn. Chalmers (1975) and Bendel and Mickey (1978) develop procedures that satisfy the definition of a correlation matrix (positive definite with 1's on the diagonal) and whose eigenvalues are approximately equal to a specified set. Another way of restricting the space from which random correlation matrices is to be drawn is to specify the expected values of the correlation matrices.

Marsaglia and Olkin (1984) consider the problem of generating correlation matrices with a given mean and the problem of generating correlation matrices with a given set of eigenvalues. Their algorithm for the latter is similar to that of Chalmers (1975). Starting with a diagonal matrix Λ with elements $0 \le \lambda_1, \lambda_2, \ldots, \lambda_d \le 1$ and such that $\sum \lambda_j = d$, the algorithm forms a random correlation matrix of the form $P\Lambda P^T$. It is shown as Algorithm 3.8.

Algorithm 3.8 Marsaglia-Olkin Method for Random Correlation Matrices with Given Eigenvalues

0. Set $E = I_d$ (the $d \times d$ identity matrix) and $k = 1$.

1. Generate an d-vector w of i.i.d. standard normal deviates, form $x = Ew$, and compute $a = \sum(1 - \lambda_i)x_i^2$.

2. Generate a d-vector z of i.i.d. standard normal deviates, form $y = Ez$, and compute $b = \sum(1 - \lambda_i)x_i y_i$, $c = \sum(1 - \lambda_i)y_i^2$, and $e^2 = b^2 - ac$.

3. If $e^2 < 0$, then go to step 2.

4. Choose a random sign, $s = -1$ or $s = 1$. Set $r = \dfrac{b + se}{a}x - y$.

5. Choose another random sign and set $p_k = \dfrac{s}{(r^T r)^{\frac{1}{2}}}w$.

6. Set $E = E - rr^T$ and set $k = k + 1$.

7. If $k < d$, then go to step 1.

8. Generate a d-vector w of i.i.d. standard normal deviates, form $x = Ew$, and set $p_d = \dfrac{x}{(x^T x)^{\frac{1}{2}}}$.

9. Construct the matrix P using the vectors p_k as its rows. Deliver $P\Lambda P^T$ as the random correlation matrix. ∎

Ryan (1980) shows how to construct fixed correlation matrices with a given structure. Although the matrices are fixed, they have applications in Monte Carlo studies for generating other deviates.

Heiberger (1978), Stewart (1980), Anderson, Olkin, and Underhill (1987), and Fang and Li (1997) give methods for generating random orthogonal matrices. The methods use reflector or rotator matrices to transform a random matrix to the form desired. The probability distribution is the Haar distribution, which is an analog of a uniform distribution for orthogonal matrices.

Heiberger (1978) discussed the use of random orthogonal matrices in generating random correlation matrices. (See Tanner and Thisted, 1982, for a correction of Heiberger's algorithm. This is the method used in the IMSL Libraries.) Stewart (1980) used random orthogonal matrices to study an estimator of the condition number of a matrix. One of the methods given by Fang and Li (1997) results in matrices with quasirandom elements (see Section 6.3, page 159).

Eigenvalues of random matrices are often of interest, especially the extreme eigenvalues. One way to generate them, of course, is to generate the random matrix and then extract the eigenvalues. Marasinghe and Kennedy (1982) describe direct methods for generation of the extreme eignevalues for certain Wishart matrices. They show that the maximum eigenvalue of a random 2×2 Wishart matrix, corresponding to a sample of size n, can be generated by the inverse CDF method:

$$w = \left(1 + \sqrt{1 - u^{2/(n-1)}}\right)/2,$$

where u is generated from $U(0, 1)$.

Using a Weibull density,

$$g(y) = (n - 3)x e^{-(n-3)y^2/2},$$

as a majorizing density, they also give the following acceptance/rejection algorithm for the 2×2 case:

1. Generate u_1 from $U(0, 1)$, and set $y = -2\log(u_1)/(n - 3)$.

2. If $y \geq 1$, go to step 1.

3. Generate u_2 from $U(0, 1)$, and if $u_1 u_2 > (1 - y)^{(n-3)/2}$, then
      go to step 1;
   otherwise
      deliver $w = (\sqrt{y} + 1)/2.$

3.2.4   Points on a Sphere

Coordinates of points uniformly distributed over the surface of a sphere (here "sphere" may mean "circle" or "hypersphere") are equivalent to independent normals scaled so as to lie on the sphere (that is, the sum of their squares is equal to the radius of the sphere). This general method is equivalent to the polar methods for normal random variables discussed previously. It can be used in any dimension.

Marsaglia (1972b) describes methods for three and four dimensions that are about twice as fast as use of normal variates. For three dimensions, u_1 and u_2 are generated independently from $U(0,1)$, and if $s_1 = u_1^2 + u_2^2 \leq 1$, then the coordinates are delivered as

$$
\begin{aligned}
x_1 &= 2u_1\sqrt{1 - s_1}, \\
x_2 &= 2u_2\sqrt{1 - s_1}, \\
x_3 &= 1 - 2s_1.
\end{aligned}
$$

For four dimensions, u_1, u_2, and s_1 are generated as above (with the same rejection step), u_3 and u_4 are generated independently from $U(0,1)$, and s_2 is formed as $s_2 = u_3^2 + u_4^2$. The same rejection step is applied, and for the accepted points the coordinates are delivered as

$$
\begin{aligned}
x_1 &= x_1 \text{ as above} \\
x_2 &= x_2 \text{ as above} \\
x_3 &= u_3\sqrt{\frac{1 - s_1}{s_2}}, \\
x_4 &= u_4\sqrt{\frac{1 - s_1}{s_2}}.
\end{aligned}
$$

The IMSL routine rnsph uses these methods for 3 or 4 dimensions and uses scaled normals for higher dimensions.

Banerjia and Dwyer (1993) consider the related problem of generating random points in a ball, which would be equivalent to generating random points on a sphere and then scaling the radius by a uniform deviate. They describe a divide-and-conquer algorithm that can be used in any dimension and that is faster than scaling normals or scaling points from Marsaglia's method, assuming that the speed of the underlying uniform generator is great relative to square root computations.

3.2.5 Two-Way Tables

Boyett (1979) and Patefield (1981) consider the problem of generating random entries in a two-way table subject to given marginal row and column totals. The distribution is uniform over the integers that yield the given totals. Boyett derives the joint distribution for the cell counts and then develops the conditional distribution for a given cell, given the counts in all previous rows and columns. ("Previous" is interpreted with respect to a traversal of the row and column indices in their natural order.) Patefield then uses the conditional expected value of a cell count to generate a random entry for each cell in turn.

Let a_{ij} for $i = 1, 2, \ldots, r$ and $j = 1, 2, \ldots, c$ be the cell count, and use the "dot notation" for summation: $a_{\bullet j}$ is the sum of the counts in the jth column, for example; and $a_{\bullet\bullet}$ is the grand total. The conditional probability that the count in the $(l,m)^{\text{th}}$ cell is a_{lm} given the counts a_{ij}, for $1 \leq i < l$ and $1 \leq j < m$ is

$$
\frac{\left(a_{l\bullet} - \sum_{j<m} a_{lj}\right)! \left(a_{\bullet\bullet} - \sum_{i\leq l} a_{i\bullet} - \sum_{j<m} a_{\bullet j} + \sum_{j<m} \sum_{i\leq l} a_{ij}\right)!}{a_{lm}! \left(a_{l\bullet} - \sum_{j\leq m} a_{lj}\right)! \left(a_{\bullet\bullet} - \sum_{i\leq l} a_{i\bullet} - \sum_{j\leq m} a_{\bullet j} + \sum_{j<m} \sum_{i\leq l} a_{ij}\right)!}
$$

$$
\times \frac{\left(a_{\bullet m} - \sum_{i\leq l} a_{im}\right)! \left(\sum_{j>m}\left(a_{\bullet j} - \sum_{i<l} a_{ij}\right)\right)!}{\left(a_{\bullet m} - \sum_{i<l} a_{im}\right)! \left(\sum_{j\geq m}\left(a_{\bullet j} - \sum_{i<l} a_{ij}\right)\right)!}
$$

For each cell a random uniform is generated and the discrete inverse CDF method is used. Sequential evaluation of this expression is fairly simple, so the probability accumulation proceeds rapidly. The full expression is evaluated only once for each cell. Patefield (1981) also speeds up the process by beginning at the conditional expected value of each cell, rather than accumulating the CDF from 0. The conditional expected value of the random count in the $(l,m)^{\text{th}}$ cell, A_{lm}, is

$$
\mathrm{E}(A_{lm} \mid a_{ij},\ 1 \leq i < l,\ 1 \leq j < m) = \frac{\left(a_{\bullet m} - \sum_{i=1}^{l-1} a_{im}\right)\left(a_{l\bullet} - \sum_{j=1}^{m-1} a_{lj}\right)}{\sum_{j=m}^{c}\left(a_{\bullet j} - \sum_{i=1}^{l-1} a_{ij}\right)},
$$

unless the denominator is 0, in which case $\mathrm{E}(A_{lm}|a_{ij})$ is zero.

Patefield (1981) gives a Fortran program implementing the method described. This is the method used in the IMSL routing rntab.

3.2.6   Other Specific Multivariate Distributions

The Dirichlet distribution is a multivariate extension of a beta distribution, and its density is the obvious extension of the beta density (equation (3.5)),

$$
p(x) = \frac{\Gamma(\Sigma\alpha_j)}{\prod \Gamma(\alpha_j)} \prod x_j^{\alpha_j - 1}(1 - x_1 - x_2 - \cdots - x_d)^{\alpha_{d+1} - 1},
$$
$$
\text{for } 0 \leq x_j \leq 1.
$$

Arnason and Baniuk (1978) consider several ways to generate deviates from the Dirichlet distribution, including a sequence of conditional betas and the use of the relationship of order statistics from a uniform distribution to a Dirichlet. The most efficient method they found was the use of a relationship between independent gamma variates and a Dirichlet variate. If $Y_1, Y_2, \ldots, Y_d, Y_{d+1}$ are independently distributed as gamma random variables with shape parameters $\alpha_1, \alpha_2, \ldots, \alpha_d, \alpha_{d+1}$ and common scale parameter, then the d-vector X with elements

$$
X_j = \frac{Y_j}{\sum_{k=1}^{d+1} Y_k}, \quad j = 1, \ldots, k,
$$

has a Dirichlet distribution with parameters $\alpha_1, \alpha_2, \ldots, \alpha_d$. This relationship yields the straightforward method of generating Dirichlets by generating gammas.

Various multivariate extensions of the stable distributions can be defined. Modarres and Nolan (1994) give a representation of a class of multivariate stable distributions in which the multivariate stable random variable is a weighted sum of a univariate stable random variable times a point on the unit sphere. The reader is referred to the paper for the description of the class of multivariate stables for which the method applies.

Only a few of the standard univariate distributions have standard multivariate extensions. Various applications often lead to different extensions, see Johnson and Kotz (1972). A multivariate exponential distribution can be defined in terms of Poisson shocks (see Marshall and Olkin, 1967), and variates can be generated from that distribution by generating univariate Poisson variates (see Dagpunar, 1988). The bivariate gamma distribution of Becker and Roux (1981) discussed in Section 2.2 is only one possibility for extending the gamma. Others, motivated by different models of applications, are discussed by Mihram and Hultquist (1967), Ronning (1977), and Ratnaparkhi (1981), for example. Ronning (1977) describes a method for generating variates from the multivariate gamma distribution he considered.

If the density of a multivariate distribution can be specified, it is usually possible to generate variates from the distribution using an acceptance/rejection method. The majorizing density can often be just the product density, that is, a multivariate density whose components are the independent univariate variables, as in the example of the bivariate gamma on page 53.

3.3   General Multivariate Distributions

Methods are available for generating multivariate distributions with various specific properties. Extensions have been given for multivariate versions of some of the general families of univariate distributions discussed on page 104. Parrish (1990) gives a method to generate random deviates from a multivariate Pearson family of distributions. Takahasi (1965) defines a multivariate extension of the Burr distributions. Generation of deviates from the multivariate Burr can be accomplished by transformations of univariate samples.

3.3.1   Distributions with Specified Correlations

Li and Hammond (1975) propose a method for a d-variate distribution with specified marginals and variance-covariance matrix. The Li-Hammond method uses the inverse CDF method to transform a d-variate normal into a multivariate distribution with specified marginals. The variance-covariance matrix of the multivariate normal is chosen so as to yield the specified variance-covariance matrix for the target distribution. The determination the variance-covariance matrix for the multivariate normal so as to yield the desired target distribution is difficult, however, and does not always yield a positive definite variance-covariance matrix for the multivariate normal. (An approximate variance-covariance or correlation matrix that is not positive definite can be a general problem in applications of multivariate simulation.)

Lurie and Goldberg (1998) modify the Li-Hammond approach by iteratively refining the correlation matrix of the underlying normal using the sample correlation matrix of the transformed variates. They begin with a fixed sample of t multivariate normals with the identity matrix as the variance-covariance. These normal vectors are first linearly transformed by the matrix $T^{(k)}$ as described on page 105 and then transformed by the inverse CDF method into a sample of t vectors with the specified marginal distributions. The correlation matrix of the transformed sample is computed and compared with the target correlation. A measure of the difference in the sample correlation matrix and the target correlation is minimized by iterations over $T^{(k)}$. A good starting point for $T^{(0)}$ is the $d \times d$ matrix that is the square root of the target correlation matrix R, that is, the Cholesky factor, so that $T^{(0)\mathrm{T}}T^{(0)} = R$.

The measure of the difference in the sample correlation matrix and the target correlation is a sum of squares, and so the minimization is a nonlinear least squares problem. The sample size t to use in the determination of the optimal transformation matrix must be chosen in advance. Obviously, t must be large enough to give some confidence that the sample correlation matrices reflect the target accurately. Because of the number of variables in the optimization problem, it is likely that t should be chosen proportional to d^2.

Once the transformation matrix is chosen, to generate a variate from the target distribution, first generate a variate from $N_d(0, I)$, then apply the linear transformation, and finally apply the inverse CDF transformation. To generate n variates from the target distribution, Lurie and Goldberg (1998) also suggest that the normals be adjusted so that the sample has a mean of 0 and a variance-covariance matrix exactly equal to the expected value that the transformation would yield. (This is *constrained sampling*, as discussed on page 138.)

Vale and Maurelli (1983) also generate general random variates using a multivariate normal distribution as a starting point for the target correlation matrix. The distributions they consider are those for which the first four marginal moments are given. Parrish (1990), as mentioned above, gives a method for generating variates from a multivariate Pearson family of distributions. A member of the Pearson family is specified by the first four moments also.

For discrete multivariate distributions, the transformations from the multivariate normal distribution are not so obvious. A multivariate Bernoulli distribution of correlated binary random variables has applications in modeling system reliability, clinical trials with repeated measures, and genetic transmission of disease. For the multivariate Bernoulli distribution with marginal probabilities $\pi_1, \pi_2, \ldots, \pi_d$ and pairwise correlations ρ_{ij}, Emrich and Piedmonte (1991) propose identifying a multivariate normal distribution with similar pairwise correlations. The normal is determined by solving for normal pairwise correlations,

r_{ij}, in a system of $d(d-1)/2$ equations involving the bivariate standard normal CDF, Φ, evaluated at percentiles z_π corresponding to the Bernoulli probabilities:

$$\Phi(z_{\pi_i}, z_{\pi_j}; r_{ij}) = \rho_{ij}\sqrt{\pi_i(1-\pi_i)\pi_j(1-\pi_j)} + \pi_i\pi_j. \tag{3.7}$$

Once these pairwise correlations are determined, a multivariate normal y is generated and transformed to a Bernoulli, x, by the rule

$$\begin{aligned} x_i &= 1 \quad \text{if } y_i \le z_{\pi_i} \\ &= 0 \quad \text{otherwise.} \end{aligned}$$

Sums of multivariate Bernoulli random variables are multivariate binomial random variables. Phenomena modeled by binomial distributions, within clusters, often exhibit greater or less intracluster variation than independent binomial distributions would indicate. This behavior is called "over-dispersion" or "under-dispersion". Over-dispersion can be simulated by the beta-binomial distribution discussed earlier. A beta-binomial cannot model under-dispersion, but the method of Emrich and Piedmonte (1991) to induce correlations in the Bernoulli variates can be used to model either over-dispersion or under-dispersion. Ahn and Chen (1995) discuss this method and compare it with the use of a beta-binomial in the case of over-dispersion. They also compared the output of the simulation models for both under-dispersed and over-dispersed binomials with actual data from animal litters.

Park, Park, and Shin (1996) give a method for generating correlated binary variates based on sums of Poisson random variables in which the sums have some terms in common. They let Z_1, Z_2, and Z_3 be independent Poisson random variables with nonnegative parameters $\alpha_{11} - \alpha_{12}$, $\alpha_{11} - \alpha_{12}$, and α_{12}, respectively, with the convention that a Poisson with parameter 0 is a degenerate random variable equal to 0, and define the random variables Y_1 and Y_2 as

$$Y_1 = Z_1 + Z_3$$

and

$$Y_2 = Z_2 + Z_3.$$

They then define the binary random variables X_1 and X_2 by

$$\begin{aligned} X_i &= 1 \quad \text{if } Y_i = 0 \\ &= 0 \quad \text{otherwise.} \end{aligned}$$

They then determine the constants, α_{11}, α_{22}, and α_{12}, so that

$$E(X_i) = \pi_i$$

and

$$\text{Corr}(X_1, X_2) = \rho_{12}.$$

It is easy to see that

$$\alpha_{ij} = \log\left(1 + \rho_{ij}\sqrt{\frac{(1 - \pi_i)(1 - \pi_j)}{\pi_i \pi_j}}\right) \qquad (3.8)$$

yields those relations. After the α's are computed, the procedure is as shown in Algorithm 3.9

Algorithm 3.9 Park/Park/Shin Method for Generating Correlated Binary Variates

0. Set $k = 0$.

1. Set $k = k + 1$. Let $\beta_k = \alpha_{rs}$ be the smallest positive α_{ij}.

2. If $\alpha_{rr} = 0$ or $\alpha_{ss} = 0$ then stop.

3. Let S_k be the set of all indices, i, j, for which $\alpha_{ij} > 0$. For all $\{i, j\} \subseteq S_k$, set $\alpha_{ij} = \alpha_{ij} - \beta_k$.

4. If not all $\alpha_{ij} = 0$ then go to step 1.

5. Generate k Poisson deviates, z_j, with parameters β_j. For $i = 1, 2, \ldots, d$, set

$$y_i = \sum_{i \in S_j} z_j.$$

6. For $i = 1, 2, \ldots, d$, set

$$\begin{aligned} x_i &= 1 \quad \text{if } y_i = 0 \\ &= 0 \quad \text{otherwise.} \end{aligned}$$

∎

Lee (1993) gives another method to generate multivariate Bernoullis using odds ratios. (Odds ratios and correlations uniquely determine each other for binary variables.)

Gange (1995) gives a method for generating more general multivariate categorical variates, using iterative proportional fitting to the marginals. Kachitvichyanukul, Cheng, and Schmeiser (1988) describe methods for inducing correlation in binomial and Poisson random variates.

3.3.2 Data-Based Random Number Generation

Other multivariate distributions can be generated in such a way as to correspond to the empirical distribution of a given sample. This kind of method is called *data-based random number generation*.

Suppose a set of observations, $\{x_1, x_2, \ldots, x_n\}$, is available, and we wish to generate a pseudorandom sample from the same distribution as the given dataset. How we proceed depends on some assumptions about the distribution. If the distribution is discrete and we have no other information about it than what is available from the given dataset, the best way of generating a pseudorandom sample from the distribution is just to generate a random sample of indices with replacement (see Chapter 4), and then to use the index set as indices for the given sample. For scalar x, this is equivalent to using the inverse CDF method on the empirical distribution,

$$P(x) = \frac{1}{n}(\text{number of } x_i \leq x).$$

This method of generating a random sample using given data is what is done in nonparametric Monte Carlo bootstrapping.

If it is assumed that the given sample is from some parametric family, one approach is to use the sample to estimate the parameters and then use one of the methods given above. This is what is done in the parametric Monte Carlo bootstrap.

If it is assumed that the distribution is continuous, but no other assumptions are made (other than perhaps some general assumptions of existence and smoothness of the probability density), the problem can be thought of as two steps. The first step is estimating the density, and the second one is to generate random deviates from that density.

For a univariate continuous distribution, a simple procedure is to form a smoothed version of the empirical distribution function, and then use the inverse CDF method. There are several ways this can be done. One way when the given sample size is small is just to connect the jump points of the empirical distribution function with line segments to form a piecewise linear, increasing function. Another way is to bin the data and form either a histogram or a frequency polygon, and then to use the inverse CDF method on the corresponding distribution function. The distribution function corresponding to a histogram is a piecewise linear function, and the distribution function corresponding to a frequency polygon is a piecewise quadratic polynomial.

As we have mentioned above, it is not practical to use the inverse CDF method directly for multivariate distributions.

Taylor and Thompson (1986) suggest a different way that avoids the step of estimating a density. The method has some of the flavor of density estimation; however, in fact it is essentially equivalent to fitting a density with a normal kernel. It uses the m nearest neighbors of a randomly selected point; m is a *smoothing parameter*, which is a term we will have occasion to use many times subsequently. The method is particularly useful for multivariate data. Suppose the given sample is $\{x_1, x_2, \ldots, x_n\}$ (the x's are vectors). A random vector deviate is generated by the steps given in Algorithm 3.10.

Algorithm 3.10 Thompson-Taylor Data-Based Simulation

1. Randomly choose a point, x_j, from the given sample.

2. Identify the m nearest neighbors of x_j (including x_j): $x_{j_1}, x_{j_2}, \ldots, x_{j_m}$; and determine their mean, $\bar{x}_j$.

3. Generate a random sample, $u_1, u_2, \ldots, u_m$, from a uniform distribution with lower bound $\frac{1}{m} - \sqrt{\frac{3(m-1)}{m^2}}$ and upper bound $\frac{1}{m} + \sqrt{\frac{3(m-1)}{m^2}}$.

4. Deliver the random variate

$$z = \sum_{k=1}^{m} u_k(x_{j_k} - \bar{x}_j) + \bar{x}_j.$$

∎

The limits of the uniform weights and the linear combination for z are chosen so that the expected value of the ith element of a random variable Z that yields z is the ith element of the sample mean of the x's, $\bar{x}_i$, that is,

$$\mathrm{E}(Z_i) = \bar{x}_i.$$

(The subscripts in these expressions refer to the elements of the data vectors, rather than to the element of the sample.) Likewise, the variance and covariance of elements of Z are close to the sample variance and covariance of the elements of the given sample. If $m = 1$, they would be exactly the same. For $m > 1$, the variance is slightly larger, because of the variation due to the random weights. The exact variance and covariance, however, depend on the distribution of the given sample because the linear combination is of nearest points. The routine **rndat** in the IMSL Libraries implements this method.

Prior to generating any pseudorandom deviates by the Thompson-Taylor method, the given dataset should be processed into a k-d-tree (see Friedman, Bentley, and Finkel, 1977) so that the nearest neighbors of a given observation can be identified quickly.

3.4 Geometric Objects

It is often of interest to generate random geometric objects, for example, in a space-filling problem. Exercise 3.7 on page 119 discusses a simple random-packing problem to fill a volume with balls of equal size. More interesting and complicated applications involve objects of variable shape such as stones to be laid down as paving or to be dispersed through a cement mixture.

Other applications are in the solution of partial differential equations and in probability density estimation, when we may wish to tessellate space using triangles, rectangles, hexagons, or other objects. Rectangles are simple tessellating objects that have similar properties in various dimensions. (In one dimension they are intervals.) Devroye, Epstein, and Sack (1993) give methods for generating random intervals and hyperrectangles.

Exercises

3.1. Polar transformations.

   (a) Show that the Box-Muller transformation (3.1), page 88, correctly transforms independent uniform variates into independent standard normals.

   (b) Let X_1 and X_2 be independent standard normal random variables and represent the pair in polar coordinates:

$$X_1 = R\cos\Theta$$
$$X_2 = R\sin\Theta.$$

   Show that R and Θ are independent, and that Θ has a $U(0, 2\pi)$ distribution and R^2 has an exponential distribution with parameter $\frac{1}{2}$; that is, $f_R(r) = re^{-r^2/2}$.

   (c) Show that the rejection polar method of Algorithm 3.1 does indeed deliver independent normals.

   (d) Write a Fortran or a C program implementing both the regular Box-Muller transformation using the intrinsic system functions and the rejection polar method of Algorithm 3.1. Empirically compare the efficiencies of the two methods by timing the execution of your program.

3.2. Truncated normal.

   (a) Write the steps for generating a truncated normal using the method of Robert (1995) (page 92).

   (b) Show that for majorizing densities that are translated exponentials, the one with scale parameter λ^* in equation (3.2) is optimal. (Optimal means it results in the greatest probability of acceptance.)

3.3. Prove that the random variable delivered in equation (3.3) on page 92 has an exponential distribution.

3.4. Identify quantities in Algorithm 3.2, page 94, that should be computed as part of a setup step prior to the steps shown. There are four quantities that should be precomputed, including such trivial new variables as $a = \alpha - 1$.

3.5. Write a program to generate truncated gamma variates using the method of Philippe (1997) (page 96). The program should accept α, β, τ, and n^* as arguments. Now, write a program that uses the Cheng/Feast (1979) algorithm, page 94, and which just rejects variates greater than τ. (Remember to precompute quantities such as that are used in every pass through the accept/reject loop.) For $\alpha = 5$, $\beta = 1$, and various values of τ, generate 10,000 deviates by both methods and compare the times of the two algorithms.

3.6. Correlated binary variates.

   (a) Derive equation (3.8) on page 115. (See Park, Park, and Shin, 1996.)

   (b) Consider the pair of binary variates with $\pi = (0.2, 0.7)$ and $\rho = 0.05$. Generate 1000 pairs using the method of Emrich and Piedmonte (1991). First compute $r = 0.0961$ from equation (3.7). Likewise, generate 1000 pairs using the method of Park, Park and Shin (1996)and compute the sample correlation. How close are they to ρ?

3.7. Consider a cube with each edge of length s. Imagine the cube to represent a box with no top. Develop a program to simulate the dropping of balls of diameter r ($r < s$) into the box until it is filled. ("Filled" can mean either that no more balls will stack on the others, or that no ball has its top higher than the top of the box. Your program should allow either definition.) Assume that the balls will roll freely to their lowest local energy state within the constraints imposed by the sides of the box and the other balls. Let $s = 10$ and $r = 3$. How many balls go into the box? Try the simulation several times. Do you get the same number? Now let $s \gg r$. How many balls go into the box? Try the simulation several times. Do you get the same number? Is this number in accordance with the Kepler conjecture? (The Kepler conjecture states that the densest arrangement of balls is obtained by stacking triangular layers of balls, the "cannonball arrangement".)

Chapter 4

Generation of Random Samples and Permutations

4.1 Random Samples

In applications of statistical techniques as well as in Monte Carlo studies it is often necessary to take a random sample from a given finite set. A common form of random sampling is simple random sampling without replacement, in which a sample of n items is selected from a population N in such a way that every subset of size n from the universe of N items has an equal chance of being the sample chosen. This is equivalent to a selection mechanism in which n different items are selected, each with equal probability, n/N, and without regard to which other items are selected.

In a variation, called Bernoulli sampling, each item is selected independently with a probability of n/N. In this variation, the sample size itself is a random variable. In another variation, called simple random sampling with replacement, a fixed sample of size n is chosen with equal and independent probabilities, without the restriction that the sample items be different. Selection of a simple random sample with replacement is equivalent to generating n random deviates from a discrete uniform distribution over the integers $1, 2, \ldots, n$.

Variations in sampling designs include stratified sampling and multistage sampling. In multistage sampling, *primary sampling units* are selected first. The primary sampling units consist of *secondary sampling units*, which are then sampled in a second stage. For example, the primary sampling units may be geographical districts, and the secondary sampling units may be farms. The more complicated sampling schemes can be implemented by using the same algorithms as for the simpler designs on different sampling units or at different stages.

Often, rather than each item in the population having the same probability of being included in the sample, the ith item in the population has a preassigned probability, p_i. In many cases, the p_i's are assigned based on some auxiliary variable, such as a measure of size of the population units. In multistage sampling, the primary sampling units are often chosen with probability proportional to the number of secondary units they contain.

In one variation of sampling with unequal probabilities, similar to Bernoulli sampling, each item in the population is selected independently with its pre-assigned probability. This is called Poisson sampling. In this case, the sample size is a random variable.

There are other operational variations. Sometimes the sampling activity begins with the generation of a set of indices, which are then used to determine whether a given member of the population is included in the population. In a more common situation, however, the items in the population are encountered sequentially and a decision must be made at the time of encountering the item. This is called a "draw sequential" scheme.

The method shown in Algorithm 4.1 is a draw sequential scheme for simple random sampling if the $p_i = n/N$ for each i. If $\sum p_i$ is known in advance (and presumably $\sum p_i = n_0$, where n_0 is some expected sample size), Algorithm 4.1 yields a probability sample without replacement, but the sample size is a random variable. Unless the sample size is fixed to be 1 or all of the p_i's are equal, the schemes to achieve a fixed sample size are very complicated. See Särndal, Swensson, and Wretman (1992) for discussions of various sampling designs.

Algorithm 4.1 Draw Sequential Probability Sampling

0. Set $i = 0$, $s_i = 0$, $t_i = 0$, and $T = \sum p_j$.

1. Set $i = i + 1$ and generate u from $U(0, 1)$.

2. If $u \leq \frac{Tp_i - s_{i-1}}{T - t_{i-1}}$, then
   include the ith item and set $s_i = s_{i-1} + p_i$.

3. If $i < N$, set $t_i = t_{i-1} + p_i$ and go to step 1.   ∎

This algorithm is obviously $O(N)$. For simple random sampling, that is, if the probabilities are all equal, we can improve on Algorithm 4.1 in two ways. If N is known in advance, Vitter (1984) and Ahrens and Dieter (1985) give methods to obtain the simple random sample in $O(n)$ time, because we can generate the random skips in the population before the next item is selected for inclusion in the sample.

The other improvement for the case of simple random sampling is to drop the requirement that N be known in advance. If a sample of size n is to be selected from a population of size N, which is not known in advance, a "reservoir sample" of size n is filled and then updated by replacements until the full population has been considered; the reservoir is the sample. This process is called reservoir sampling. In a simple-minded implementation of a reservoir algorithm, a decision is made in turn, for each population item, whether to replace one of the reservoir items with it. This process is obviously $O(N)$

(McLeod and Bellhouse, 1983). This can also be improved on by generating a random number of population items to skip before including one in the sample.

Any one-pass algorithm for the case when N is unknown must be a reservoir method (Vitter, 1985). Li (1994) summarizes work on reservoir sampling and gives the following algorithm, which is $O(n(1 + \log(N/n)))$. In this description, let the population items be denoted by $X_1, X_2, X_3, \ldots$.

Algorithm 4.2 Li's Reservoir Sampling

    0. Initialize the reservoir with the first n items of the population. Call the sample items $x_1, x_2, \ldots, x_n$. Set $i = n + 1$.

    1. Generate u from $U(0, 1)$ and compute $w = u^{\frac{1}{n}}$.

    2. Generate u from $U(0, 1)$ and compute $s = \left\lfloor \frac{\log u}{\log(1 - w)} \right\rfloor$.

    3. If X_{i+s} is in the population, then
        3.a. generate j uniformly from $1, 2, \ldots, n$; replace x_j with X_{i+s};
           and go to step 1;
    otherwise
        3.b. terminate. ∎

At first, one may think that it would be unreasonable to choose an n for a sample size without knowing N; but indeed, this is a common situation, especially in sampling for proportions. Using the normal approximation for setting a confidence interval, a sample of size 1067 will yield a 95% confidence interval no wider than 0.06 for a population proportion, even if the population correction factor is ignored. In practice, a "random sample" of size 1100 is often used in sampling for proportions, no matter what the size of the population. Often, of course, simple random samples are not used. Instead, systematic samples or cluster samples are used because of the operational simplicity.

Because databases are often organized in computer storage as trees rather than just as sequential files, it is sometimes necessary to be able to sample from trees. Knuth (1975) gives a general sampling algorithm for trees and Rosenbaum (1993) describes a more efficient method for equal probability sampling from trees. Alonso and Schott (1995) and Wilson and Propp (1996) describe algorithms for generating random spanning trees from a directed graph. Olken and Rotem (1995a) give a survey of methods for sampling from databases organized as trees or with more complex structures.

Olken and Rotem (1995b) extend the reservoir sampling methods to the problem of sampling from a spatial database such as may be used in geographic information systems (GIS).

4.2 Permutations

The problem of generation of a random permutation is related to random sampling. A random permutation can be generated in one pass as shown in Algorithm 4.3. The algorithm is $O(N)$. In this description, let the population items be indexed by $1, 2, 3, \ldots N$.

Algorithm 4.3 Random Permutation

0. Set $i = 0$.

1. Generate j uniformly from the integers $1, 2, \ldots, N - i$.

2. Exchange the elements in the $(N - i)^{\text{th}}$ and the jth positions.

3. If $i < N - 2$, then
   3.a. set $i = i + 1$ and go to step 1;
   otherwise
   3.b. terminate.

Limitations of Random Number Generators

We have mentioned possible problems with random number generators that have short periods. As we have seen, the period of the most commonly used generators are of the order of 2^{31} (or 10^9), although some generators may have a substantially longer period. We have indicated that a long period is desirable, and this is intuitively obvious. Generation of random samples and permutations is a simple problem that really points out a limitation of a finite period.

Both of these problems involve a classic operation for yielding large numbers: the factorial operation. Consider the seemingly simple case of generating a random sample of size 500 from a population of size 1,000,000. The cardinality of the sample space is

$$\binom{1000000}{500} \approx 10^{2000}.$$

This number is truly large by almost any scale. It is clear that no practical finite period generator can cover this sample space. (See Greenwood, 1976a, 1976b.)

In no other application considered in this book does the limitation of a finite period stand out so clearly as it does in the case of generation of random samples and permutations.

4.3 Generation of Nonindependent Samples

A sequence of random variables can be thought of either as a stochastic process or as a single multivariate random variable. Many multivariate distributions can usefully be formulated as stochastic processes, and the conditional generation algorithms we described for the multivariate normal and the multinomial distributions are simple examples of this. General methods for inducing correlations either within a single stream of random variates or between two streams have been described by Kachitvichyanukul, Cheng, and Schmeiser (1988) and by Avramidis and Wilson (1995).

4.3.1   Order Statistics

In some applications a random sample needs to be sorted into ascending or descending order, and in some cases we are interested only in the extreme values of a random set. In either case, the random numbers are *order statistics* from a random sample. In the early 1970s, several authors pointed out that order statistics could be generated directly, without simulating the full set and then sorting (Lurie and Hartley, 1972; Reeder, 1972; and Schucany, 1972). There are three ways of generating order statistics, other than just generating the variates and sorting them:

1. *Sequential*: Generate one order statistic directly, using its distribution, and then generate additional ones sequentially, using their conditional distributions.

2. *Spacings*: Generate one order statistic directly, using its distribution, and then generate additional ones sequentially, using the distribution of the spacings between the order statistics.

3. *Multinomial groups*: Divide the range of the random variable into intervals, generate one multinomial variate for the counts in the intervals corresponding to the probability of the interval, and then generate the required number of ordered variates within each interval by any of the previous methods.

These methods, of course, apply to any univariate distribution. If the target distribution is not the uniform distribution, one of these methods could be applied to the target distribution directly. If the inverse CDF method can be applied, however, it is usually better to generate the appropriate order statistics from the uniform distribution and then use the inverse CDF method to generate the order statistics from the target distribution.

Reeder (1972), Lurie and Mason (1973), and Rabinowitz and Berenson (1974) compare the performance of these methods for generating order statistics from a uniform distribution, and Gerontidis and Smith (1982) compare the performance for some nonuniform distributions. If all n order statistics are required, the multinomial groups method is usually best. Often, however, only a few order statistics from the full set of n are required. In that case either the sequential method or the spacings method is to be preferred.

It is easy to show that the ith order statistic from a sample of size n from a $U(0,1)$ distribution has a beta distribution with parameters i and $n-i+1$ (see, e.g., David, 1981). Using this fact for the uniform distribution, we see how to start the sequential and the spacings methods described above.

Notice that for the first or the nth order statistic, the beta distribution has one parameter equal to 1. For such a beta distribution, it is easy to generate a deviate, because the density, and hence the distribution function, is proportional to a monomial. The inverse CDF method just involves taking the nth root of a uniform.

To proceed with the sequential method, we use the fact that conditional on $u_{(i)}$, for $j > i$, the jth order statistic from a sample of size n from a $U(0,1)$ distribution has the same distribution as the $(j-i)^{\text{th}}$ order statistic from a sample of size $n-i$ from a $U(u_{(i)}, 1)$ distribution. We again use the relationship to the beta distribution, except we have to scale it into the interval $(u_{(i)}, 1)$.

The spacings method depends on an interesting relationship between ratios of independent standard exponential random variables and the difference between order statistics from a $U(0,1)$ distribution. It is easy to show that if $Y_1, Y_2, \ldots, Y_n, Y_{n+1}$ are independently distributed as exponential with parameter equal to 1, for $i = 1, 2, \ldots, n$, the ratio

$$R_i = \frac{Y_i}{Y_1 + Y_2 + \ldots + Y_n + Y_{n+1}}$$

has the same distribution as

$$U_{(i)} - U_{(i-1)}$$

where $U_{(i)}$ is the ith order statistic from a sample of size n from a $U(0,1)$ distribution, and $U_{(0)} = 0$.

Nagaraja (1979) points out an interesting anomaly: although the distribution of the order statistics using any of three methods we have discussed is the same,

$$\Pr\left(X_{(n)}^{\text{seq}} > X_{(n)}^{\text{sort}}\right) > 0.5,$$

where $X_{(n)}^{\text{seq}}$ is the maximum uniform order statistic generated by first generating $X_{(1)}$ as the nth root of a uniform and then generating the sequence up to $X_{(n)}$, and $X_{(n)}^{\text{sort}}$ is the maximum uniform order statistic generated by sorting. The probability difference is small, and this fact is generally not important in Monte Carlo studies.

4.3.2 Nonindependent Sequences: Nonhomogeneous Poisson Process

Some important Poisson processes have interarrival times that are distributed as exponential with rate parameter λ varying in time, $\lambda(t)$. For a nonhomogeneous Poisson process with a rate function that is varying in time, Lewis and Shedler (1979) develop a method called "thinning" that uses acceptance/rejection on Poisson variates from a process with a greater rate function, λ^* (which may be stepwise constant). The method is shown in Algorithm 4.4.

Algorithm 4.4 Nonhomogeneous Poisson Process by Thinning

0. Set $i = 0$ and set $t_i = t_0$.

1. Set $d = 0$.

2. Generate e from an exponential with rate function λ^*, and set $d = d + e$.

3. Generate a u from $U(0,1)$.

4. If $u \leq \lambda(t)/\lambda^*$, then set $i = i + 1$, set $t_i = t_{i-1} + d$, deliver t_i, and go to step 1.

5. Go to step 2.   ∎

The IMSL Library routine **rnnpp** generates a nonhomogeneous Poisson process using the Lewis and Shedler thinning method.

The thinning technique also applies to other distributions that are varying in time. Chen and Schmeiser (1992) describe methods for generating a nonhomogeneous Poisson process in which the rate varies cyclically as a trigonometric function.

4.3.3 Censored Data

In applications studying lifetime data, such as reliability studies and survival analysis, it is common not to observe all lifetimes of the experimental units in the study. The unobserved data are said to be "censored". There are several ways the censoring can be performed. Often, either the very small values or the very large values are not observed. Data collected by such procedures can be simulated as truncated distributions, such as we have already discussed.

More complicated censoring may be difficult to simulate. Balakrishnan and Sandhu (1995) give a method for progressive Type-II censoring. In this kind of censoring, n items begin in the study. At the ith failure, r_i items are removed from the test; that is, their lifetimes are censored. In this type of accelerated life testing, we observe m failures (or lifetimes), and

$$n = m + r_1 + \cdots + r_m.$$

Balakrishnan and Sandhu (1995) show that if $U_{(1)}, U_{(2)}, \ldots, U_{(m)}$ are progressive Type-II censored observations from $U(0,1)$,

$$V_i = \frac{1 - U_{(m-i+1)}}{1 - U_{(m-i)}}, \quad \text{for } i = 1, 2, \ldots, m-1,$$

and

$$V_m = 1 - U_{(1)},$$

and

$$W_i = V_i^{i + r_m + r_{m-1} + \cdots + r_{m-i+1}}, \quad \text{for } i = 1, 2, \ldots, m,$$

then the W_i are i.i.d. $U(0,1)$.

This fact (reminiscent of the relationships of order statistics from the uniform distribution), provides the basis for simulating progressive Type-II censored observations. Also similar to the situation for order statistics, the uniform case can yield samples from other distributions by use of the inverse CDF method.

Algorithm 4.5 Progressive Type-II Censored Observations from a Distribution with CDF P

1. Generate m independent $U(0,1)$ deviates, w_i.

2. Set $v_i = w_i^{1/(i+r_m+r_{m-1}+\cdots+r_{m-i+1})}$.

3. Set $u_i = 1 - v_m v_{m-1} v_{m-i+1}$.
   The u_i are deviates from progressive Type-II censoring from the $U(0,1)$ distribution.

4. Set $x_i = P^{-1}(u_i)$ (or use the appropriate modification discussed in Chapter 2 if P is not invertible). ∎

Exercises

4.1. One scheme for sampling with replacement and with probability proportional to size is called Lahiri's method (see Särndal, Swensson, and Wretman, 1992). Suppose a population consists of N items, with sizes $M_1, M_2, \ldots, M_N$, and a sample of size n is to be selected with replacement in such a way that the probability of the ith item is proportional to M_i. (The N population items may be primary sampling units and the M_i's may be numbers of secondary sampling units in the primary units.) Lahiri's method uses acceptance/rejection:

(0) Set $i = 0$, $j = 0$, and $M = \max(M_1, M_2, \ldots, M_N)$.

(1) Set $i = i+1$ and generate k from a discrete uniform over $1, 2, \ldots, N$.

(2) Generate m from a discrete uniform over $1, 2, \ldots, M$.

(3) If $m \leq M_k$, then
     include the kth item and set $j = j + 1$.

(4) If $j < n$, go to step 1.

Prove that this method results in a probability proportional-to-size sample with replacement.

4.2. Consider the problem of generating a simple random sample of size n from a population of size N. Suppose $N = 10^6$, as in the discussion on page 124, and suppose you have a random number generator with period 10^9.

(a) With this generator, what is the largest sample you can draw in accordance with the definition of a simple random sample?

(b) Suppose you wish to use the generator to draw a random sample of size 500. How serious do you believe the problem to be? Discuss.

4.3. Prove that the ith order statistic from a sample of size n from a $U(0,1)$ distribution has a beta distribution with parameters i and $n-i+1$. *Hint:* Just write out the density as in David (1981), for example, plug in the distribution function of the uniform, and identify the terms.

4.4. Write a Fortran or C function to accept a, b, n, n_1, and n_2 (with $1 \leq n_1 \leq n_2 \leq n$), and to return the n_1^{th} through the n_2^{th} order statistics from a two-parameter Weibull distribution with parameters a and b. Use the random spacings method to generate uniform order statistics and then use the inverse CDF to generate order statistics from a Weibull distribution.

4.5. Derive two different algorithms and write programs implementing them to generate a deal of hands for the game of bridge. (That is, form four random mutually exclusive sets each containing 13 of the integers $1, 2, \ldots, 52$.) Base one algorithm on the techniques of random sampling and the other on methods of random permutations. Which is more efficient?

Chapter 5

Monte Carlo Methods

The most common applications of Monte Carlo methods in numerical computations are for evaluating integrals. Monte Carlo methods can also be used in solving systems of equations (see Chapter 7 of Hammersley and Handscomb, 1964, for example), but other methods are generally better, especially for matrices that are not sparse.

5.1 Evaluating an Integral

Monte Carlo simulation is the use of experiments with random numbers to evaluate a mathematical expression. In its simplest form (and all instances of Monte Carlo simulation can be reduced to this form), it is the evaluation of a definite integral

$$\theta = \int_a^b f(x)\,\mathrm{d}x \qquad (5.1)$$

by identifying a random variable Y with support on (a, b) and density $p(y)$, and a function g such that the expected value of $g(Y)$ is θ:

$$
\begin{aligned}
\mathrm{E}(g(Y)) &= \int_a^b g(y)p(y)\,\mathrm{d}y \\
&= \int_a^b f(y)\,\mathrm{d}y \\
&= \theta.
\end{aligned}
$$

In many cases Y is taken to be a random variable with a uniform density over the interval $[a, b]$ and g is taken to be f; hence,

$$\theta = (b-a)\mathrm{E}(f(Y)).$$

The problem of evaluating the integral becomes the familiar statistical problem of estimating a mean, $\mathrm{E}(f(Y))$.

The statistician quite naturally takes a random sample and uses the sample mean: an estimate of θ is

$$\widehat{\theta} = (b - a)\frac{\sum f(y_i)}{n}, \tag{5.2}$$

where the y_i are values of a random sample of size n from a uniform distribution over (a, b). The estimate is unbiased:

$$
\begin{aligned}
\mathrm{E}(\widehat{\theta}) &= (b - a)\frac{\sum \mathrm{E}(f(Y_i))}{n} \\
&= (b - a)\mathrm{E}(f(Y)) \\
&= \int_a^b f(x)\,\mathrm{d}x.
\end{aligned}
$$

The variance is

$$
\begin{aligned}
\mathrm{V}(\widehat{\theta}) &= (b - a)^2\frac{\sum \mathrm{V}(f(Y_i))}{n^2} \\
&= \frac{(b - a)^2}{n}\mathrm{V}(f(Y)) \\
&= \frac{(b - a)}{n}\int_a^b \left(f(x) - \int_a^b f(t)\mathrm{d}t\right)^2 \mathrm{d}x. \tag{5.3}
\end{aligned}
$$

The integral in (5.3) is a measure of the *roughness* of the function. (There are various ways of defining roughness. Most definitions involve derivatives. The more derivatives that exist, the less rough the function. Other definitions, such as the one here, are based on a norm of a function. The L_2 norm of the difference of the function from its integrated value is a very natural measure of roughness of the function. Another measure is just the L_2 norm of the function itself, which, of course, is not translation invariant.)

Following standard practice, we could use the variance of the Monte Carlo estimator to form an approximate confidence interval for the integral being estimated. Of course, the confidence limits would include the unknown terms in the variance. We could, however, estimate the variance of the estimator using the same sample that we use to estimate the integral.

The standard deviation in the approximate confidence limits is sometimes called a "probabilistic error bound". The word "bound" is misused here, of course, but in any event, the standard deviation does provide some measure of a sampling "error". The important thing to note from equation (5.3) is the order of error; it is $O(n^{-\frac{1}{2}})$.

An important property of the standard deviation of a Monte Carlo estimate of a definite integral is that the order in terms of the number of function evaluations is independent of the dimensionality of the integral. On the other hand, the usual error bounds for numerical quadrature are $O(n^{-\frac{2}{d}})$, where d is the dimensionality.

We should be aware of a very important aspect of this discussion of error bounds for the Monte Carlo estimator. It applies to random numbers. The pseudorandom numbers we actually use only simulate the random numbers, so "unbiasedness" and "variance" must be interpreted carefully.

The method of estimating an integral just described is sometimes called "crude Monte Carlo". In Exercise 5.2, page 146, we describe another method, called "hit-or-miss", and ask the reader to show that the crude method is superior to hit-or-miss.

Suppose the original integral can be written as

$$\theta = \int_a^b f(x)\mathrm{d}x$$
$$= \int_a^b g(x)w(x)\mathrm{d}x, \tag{5.4}$$

where $w(x)$ is a probability density over (a, b). (As with the uniform example above, it may require some scaling to get the density to be over the interval (a, b).) Now, suppose we can generate random variates y_i from the distribution with density w. Then our estimate of θ is just

$$\widehat{\theta} = \frac{\sum g(y_i)}{n}. \tag{5.5}$$

The variance of the estimator in (5.5) is likely to be smaller than that of the estimator in (5.2) (see Exercise 5.6 on page 149 and see Section 5.3). The use of a probability density as a weighting function also allows us to apply the Monte Carlo method to improper integrals (see Exercise 5.7).

The Monte Carlo quadrature methods extend directly to multivariate integrals, although, obviously, it takes larger samples to fill the space. It is, in fact, only for multivariate integrals that Monte Carlo quadrature should ordinarily be used. The preference for Monte Carlo in multivariate quadrature results from the independence of the pseudoprobabilistic error bounds and the dimensionality mentioned above.

5.2 Variance of Monte Carlo Estimators

Monte Carlo methods are sampling methods; therefore the estimates that result from Monte Carlo procedures have associated *sampling errors*. The fact that the estimate is not equal to its expected value (assuming the estimator is unbiased) is not an "error" or a "mistake"; it is just a result of the variance of the random (or pseudorandom) data.

As in any statistical estimation problem, an estimate should be accompanied by an estimate of its variance. The estimate of the variance of the estimator of interest is usually just the sample variance of computed values of the estimator of interest. A Monte Carlo estimate usually has the form of the estimator of θ in equation (5.2):

$$\widehat{\theta} = c\,\frac{\sum f_i}{n},$$

for which a variance estimator is

$$\widehat{V}(\widehat{\theta}) = c^2\,\frac{\sum(f_i - \bar{f})^2}{n-1}. \tag{5.6}$$

This is because the elements of the set of random variables $\{F_i\}$, on which we have observations $\{f_i\}$, are (assumed to be) independent.

If the F_i do not have zero correlations, the estimator (5.6) has an expected value that includes the correlations; that is, it is biased for estimating $V(\widehat{\theta})$. This situation arises often in simulation. In many processes of interest, however, observations are "more independent" of observations farther removed within the sequence than they are of observations closer to them in the sequence. A common method for estimating the variance in a sequence of nonindependent observations, therefore, is to use the means of successive subsequences that are long enough that the observations in one subsequence are almost independent of the observations in another subsequence. The means of the subsequences are called "batch means".

If $F_1, F_2, \ldots, F_m, F_{m+1}, \ldots, F_{2m}, F_{2m+1}, \ldots, F_{km}$ is a sequence of random variables such that the correlation of F_i and F_{i+m} is approximately zero, an estimate of the variance of the mean, $\bar{F}$, of the $n = km$ random variables can be developed by observing that

$$
\begin{aligned}
V(\bar{F}) &= V\left(\frac{1}{n}\sum F_i\right) \\
&= V\left(\frac{1}{k}\sum_{j=1}^{k}\left(\frac{1}{m}\sum_{i=(j-1)m+1}^{jm} F_i\right)\right) \\
&\approx \frac{1}{k^2}\sum_{j=1}^{k} V\left(\frac{1}{m}\sum_{i=(j-1)m+1}^{jm} F_i\right) \\
&\approx \frac{1}{k} V(\bar{F}_m),
\end{aligned}
$$

where $\bar{F}_m$ is the mean of a batch of length m. If the batches are long enough, it may be reasonable to assume the means have a common variance. An estimator of the variance of $\bar{F}_m$ is the standard sample variance from k observations, $\bar{f}_1, \bar{f}_2, \ldots, \bar{f}_k$:

$$\frac{\sum(\bar{f}_j - \bar{f})^2}{k-1}.$$

Hence, the batch-means estimator of the variance of $\bar{F}$ is

$$\widehat{V}(\bar{F}) = \frac{\sum(\bar{f}_j - \bar{f})^2}{k(k-1)}. \tag{5.7}$$

This batch-means variance estimator should be used if the Monte Carlo study yields a stream of nonindependent observations, such as in a time series or when the simulation uses a Markov chain. The size of the subsamples should be as small as possible and still have means that are independent. A test of the independence of the $\bar{F}_m$ may be appropriate to help in choosing the size of the batches. Batch means are useful in variance estimation whenever a Markov chain is used in the generation of the random deviates, as discussed in Section 2.7.

5.3 Variance Reduction

An objective in sampling is to reduce the variance of the estimators while preserving other good qualities, such as unbiasedness. In this section we briefly discuss variance reduction in Monte Carlo applications. The emphasis on efficient Monte Carlo sampling goes back to the early days of digital computing (Kahn and Marshall, 1953); but the issues are just as important today (or tomorrow), because, presumably, we are solving bigger problems.

Except for straightforward analytic reduction, discussed in the next section, techniques for reducing the variance of a Monte Carlo estimator are called "swindles" (especially if they are thought to be particularly clever).

5.3.1 Analytic Reduction

The first principle in estimation is to use any known quantity to improve the estimate. For example, suppose the problem is to evaluate the integral

$$\theta = \int_a^b f(x)\,\mathrm{d}x$$

by Monte Carlo. Consider the representation of the integral:

$$\theta = \int_a^c f(x)\,\mathrm{d}x + \int_c^b f(x)\,\mathrm{d}x$$

$$= \theta_1 + \theta_2.$$

Now suppose a part of this decomposition of the original problem is known, that is, suppose we know θ_1. If $a < b < c$, it is almost certain that it would be better to use Monte Carlo only to estimate θ_2, and take as our estimate of θ the sum of the known θ_1 and the estimated value of θ_2. This seems intuitively obvious, and it is true unless there is some relationship between $f(x)$ and $f(x+t)$, where x is in (a, c) and $x + t$ is in (c, b). If there is some known relationship, it may be possible to improve the estimate $\widehat{\theta}_2$ of θ_2 by using a translation of the same random numbers used for $\widehat{\theta}_1$ to estimate θ_1. For example, if $\widehat{\theta}_1$ is larger than the known value of θ_1, the proportionality of the overestimate, $(\widehat{\theta}_1 - \theta_1)/\theta_1$, may be used to adjust $\widehat{\theta}_2$. This is the same principle as ratio or regression estimation in ordinary sampling theory (see, for example, Särndal, Swensson, and Wretman, 1992).

Now consider a different representation of the integral, in which f is expressed as $g + h$:

$$\theta = \int_a^b (g(x) + h(x))\, \mathrm{d}x$$

$$= \int_a^b g(x)\, \mathrm{d}x \; + \; \int_a^b h(x)\, \mathrm{d}x$$

$$= \theta_3 \qquad\qquad + \quad \theta_4,$$

and suppose a part of this decomposition, say θ_3, is known. In this case, the use of the known value of $\int_a^b g(x)\, \mathrm{d}x$ is likely to help only if $g(x)$ tends to vary similarly with $h(x)$. In this case it would be better to use Monte Carlo only to estimate θ_4, and take as our estimate of θ the sum of the known θ_3 and the estimated value of θ_4. This is because $h(x) - g(x)$ is less rough than $f(x)$. Also, as in the case above, if there is some known relationship between $g(x)$ and $h(x)$, such as one tends to decrease as the other increases, it may be possible to use the negative correlation of the individual estimates to reduce the variance of the overall estimate.

5.3.2   Antithetic Variates

Again consider the problem of estimating the integral

$$\theta = \int_a^b f(x)\, \mathrm{d}x$$

by Monte Carlo. The standard crude Monte Carlo estimator, equation (5.2), is $(b - a) \sum f(x_i)/n$, where x_i is uniform over (a, b). It would seem intuitively plausible that our estimate would be subject to less sampling variability if, for each x_i, we used its "mirror"

$$x_i' = a + (b - x_i).$$

This mirror value is called an antithetic variate, and use of antithetic variates can be effective in reducing the variance of the Monte Carlo estimate, especially if the integral is nearly uniform. For a sample of size n, the estimator is

$$\frac{b - a}{n} \sum_{i=1}^{\frac{n}{2}} (f(x_i) + f(x_i'))$$

The variance of the sum is the sum of the variances plus twice the covariance. Antithetic variates have negative covariances, thus reducing the variance of the sum.

Antithetic variates from distributions other than the uniform can also be formed. The linear transformation that works for uniform antithetic variates cannot be used, however. A simple way of obtaining negatively correlated variates from other distributions is just to use antithetic uniforms in the inverse CDF. Schmeiser and Kachitvichyanukul (1990) discuss other ways of doing this. If the variates are generated using acceptance/rejection, for example, antithetic variates can be used in the majorizing distribution.

5.3.3 Importance and Stratified Sampling

If the integral being evaluated is not nearly uniform, a better way of reducing the variance is by sampling more heavily in some regions than in others. In importance sampling, regions corresponding to large values of the integrand are sampled more heavily, as would be done in equation (5.4) by the weighting function.

Importance sampling is similar to hit-or-miss Monte Carlo (see Exercise 5.2, page 146), and the relationship is particularly apparent when the weighting function is sampled by acceptance/rejection. (See Caflisch and Moskowitz, 1995, for some variations of the acceptance/rejection method for this application. Their variations allow a weighting of the acceptance/rejection decision, called smoothed acceptance/rejection, that appears to be particularly useful if the sampling is done from a quasirandom sequence discussed, as in Section 6.3.) In some of the literature on Monte Carlo methods, importance sampling is called use of "weight windows".

In stratified sampling, the rule is to sample where the function is rough, that is, where the values $f(x_i)$ are likely to exhibit a lot of variability. Whereas the principle of importance sampling is generally implemented continuously by using a weighting density, stratified sampling is usually done by forming distinct subregions. This is exactly the same principle applied in sampling theory in which the allocation is proportional to the variance (see Särndal, Swensson, and Wretman, 1992). In some of the literature on Monte Carlo methods, stratified sampling is called "geometric splitting".

5.3.4 Common Variates

Often in Monte Carlo simulation the objective is to estimate the differences in parameters of two random processes. The two parameters are likely to be positively correlated. If that is the case, then the variance in the individual differences is likely to be smaller than the variance of the difference of the overall estimates.

Suppose, for example, that we have two statistics, S and T, that are unbiased estimators of some parameter of a given distribution. We would like to know the difference in the variances of these estimators

$$\mathrm{V}(T) - \mathrm{V}(S)$$

(because the one with the smaller variance is better). We assume each statistic is a function of a random sample: $\{x_1, x_2, \ldots, x_n\}$. A Monte Carlo estimate of the variance of the statistic T for a sample of size n is obtained by generating m samples of size n from the given distribution, computing T_i for the ith sample, and then computing

$$\widehat{V}(T) = \frac{\sum_{i=1}^{m}(T_i - \bar{T})^2}{m - 1}.$$

Rather than doing this for T and S separately, using the unbiasedness, we could first observe

$$
\begin{aligned}
V(T) - V(U) &= E(T^2) - E(S^2) \\
&= E(T^2 - S^2), \qquad\qquad (5.8)
\end{aligned}
$$

and hence estimate the latter quantity. Because the estimators are likely to be positively correlated, the variance of the Monte Carlo estimator $\widehat{E}(T^2 - S^2)$ is likely to be smaller than the variance of $\widehat{V}(T) - \widehat{V}(S)$. If we compute $T^2 - S^2$ from each sample, that is, if we use *common variates*, we are likely to have a more precise estimate of the difference in the variances of the two estimators, T and S.

5.3.5   Constrained Sampling

Sometimes in Monte Carlo methods it is desirable that certain sample statistics match the population parameters exactly; for example, the sample mean may be adjusted by transforming each observation in the Monte Carlo sample by

$$x'_i = x_i + \mu - \bar{x},$$

where μ is the mean of the target population and $\bar{x}$ is the mean of the original Monte Carlo sample.

This idea can be extended to more than one sample statistic, but requires more algebra to match up several statistics with the corresponding parameters. Pullin (1979) describes the obvious transformations for univariate normal deviates so that the sample has a specified mean and variance, and Cheng (1985) gives an extension for the multivariate normal distribution, so that the samples generated have a specified mean and variance-covariance matrix.

Variance estimators that result from constrained samples must be used with care. This is because the constraints change the sampling variability, usually by reducing it.

5.3.6   Latin Hypercube Sampling

The techniques of sampling theory are generally designed for reducing sampling variance for single variables, often using one or just a few covariates. The statistical developments in the design of experiments provide a more powerful set of tools for reducing the variance in cases where several factors are to be investigated simultaneously. Such techniques as balanced or partially balanced fractional factorial designs allow the study of a large number of factors while keeping the total experiment size manageably small. Some processes are so complex that even with efficient statistical designs, experiments to study the process would involve a prohibitively large number of factors and levels. For some processes it may not be possible to apply the treatments whose effects are to be studied, and data are available only from observational studies. The various processes determining weather are examples of phenomena that are not amenable to traditional experimental study. Such processes can often be modeled and studied by computer experiments.

There are some special concerns in experimentation using the computer (see Sacks et al., 1989), but the issues of statistical efficiency remain. Rather than a model involving ordinary experimental units, a computer experimental model receives a fixed input and produces a deterministic output. An objective in computer experimentation (just as in any experimentation) is to provide a set of inputs that effectively (or randomly) span a space of interest. McKay, Conover, and Beckman (1979) introduce a technique called Latin hypercube sampling (as a generalization of the ideas of a Latin square design) for providing input to a computer experiment.

If each of k factors in an experiment is associated with a random input that is initially a $U(0,1)$ variate, a sample of size n that efficiently covers the factor space can be formed by selecting the ith realization of the jth factor as

$$v_j = \frac{\pi_j(i) - 1 + u_j}{n},$$

where

- $\pi_1(\cdot), \pi_2(\cdot), \ldots, \pi_k(\cdot)$ are permutations of the integers $1, \ldots, n$, sampled randomly, independently, and with replacement from the set of $n!$ possible permutations; and $\pi_j(i)$ is the ith element of the jth permutation.

- The u_j are sampled independently from $U(0,1)$.

It is easy to see that v_j are independent $U(0,1)$. We can see heuristically that such numbers tend to be "spread out" over the space.

Expressing the Monte Carlo problem as a multidimensional integral, and considering residuals from lower dimensional integrals, Stein (1987) shows that the variance of Monte Carlo estimators using Latin hypercube sampling is asymptotically smaller than the variance of Monte Carlo estimators using unrestricted random sampling. Stein's approach was to decompose the variance of the estimator of the integral into a mean, main effects of the integrated residuals of the lower dimensional integrals, and an additional residual.

Beckman and McKay (1987) and Tang (1993) provide empirical evidence that its performance is superior to simple random sampling. Owen (1992a, 1992b, 1994a), Tang (1993), and Avramidis and Wilson (1995) derive various properties of estimators based on Latin hypercube samples.

Using Stein's decomposition of the variance, Owen (1994b) showed how the variance could be reduced controlling correlations in the Latin hypercube samples.

5.4   Computer Experiments

Some of the most important questions in science and industry involve the relationship of an entity of interest to other entities that can either be controlled or more easily measured than the quantity of interest. We envision a relationship expressed by a model

$$y \approx f(x).$$

The quantity of interest y, usually called a "response" (although it may not be a response to any of the other entities), may be the growth of a crystal, the growth of a tumor, the growth of corn, the price of a stock one month hence, etc. The other variables x, called "factors", "regressors", or just "input variables", may be temperature, pressure, amount of a drug, amount of a type of fertilizer, interest rates, etc. Both y and x may be vectors. An objective is to determine a suitable form of f and the nature of the approximation. The simplest type of approximation is one in which an additive deviation can be identified with a random variable:

$$Y = f(x) + \mathrm{E}.$$

The most important objective, whatever the nature of the approximation, usually is to determine values of x that are associated with optimal realizations of Y. The association may or may not be one of causation.

One of the major contributions of the science of statistics to the scientific method is the experimental methods that efficiently help to determine f, the nature of an unexplainable deviation E, and the values of x that yield optimal values of y. Design and analysis of experiments is a fairly mature subdiscipline of statistics.

In computer experiments the function f is a computer program, x is the input, and y is the output. The program implements known or supposed relationships among the phenomena of interest. In cases of practical interest, the function is very complicated, the number of input variables may be in the hundreds, and the output may consist of many elements. The objective is to find a tractable function, $\widehat{f}$, that approximates the true behavior, at least over ranges of interest, and to find the values of the input, say $\widehat{x}_0$, such that $\widehat{f}(\widehat{x}_0)$ is optimal. How useful $\widehat{x}_0$ is depends on how close $\widehat{f}(\widehat{x}_0)$ is to $f(x_0)$, where x_0 yields the optimal value of f.

What makes this an unusual statistical problem is that the relationships are deterministic. The statistical approach to computer experiments introduces randomness into the problem. The estimate $\widehat{f}(\widehat{x}_0)$ can then be described in terms of probabilities or variances.

In a Bayesian approach, randomness is introduced by considering the function f to be a realization of a random function, F. The prior on F may be specified only at certain points, say $F(x_0)$. A set of input vectors $x_1, x_2, \ldots, x_n$ is chosen, and the output $y_i = f(x_i)$ is used to estimate a posterior distribution for $F(x)$, or at least for $F(x_0)$. See Sacks et al. (1989), Koehler and Owen

(1996), or Currin et al. (1991) for descriptions of this approach. The Bayesian approach generally involves extensive computations.

In a frequentist approach, randomness is introduced by taking random values of the input, $x_1, x_2, \ldots, x_n$. This randomness in the input yields randomness in the output $y_i = f(x_i)$, which is used to obtain the estimates $\widehat{x}_0$ and $\widehat{f}(\widehat{x}_0)$ and estimates of the variances of the estimators. See Koehler and Owen (1996) for further discussion of this approach.

5.5 Computational Statistics

The field of computational statistics includes computationally intensive statistical methods and the supporting theory. Many of these methods involve the use of Monte Carlo methods, in some cases methods that simulate a hypothesis and in other cases methods that involve resampling from a given sample or multiple partitioning of a given sample.

Some of the methods for dealing with missing data, for example, are computationally intensive, and may involve Monte Carlo resampling. An extensive discussion of these methods is given by Schafer (1997).

5.5.1 Monte Carlo Methods for Inference

Barnard (1963), in a discussion of a paper read before the Royal Statistical Society, suggested that a test statistic be evaluated for simulated random samples to determine the significance level of the test statistic computed from a given dataset. This kind of procedure is called a Monte Carlo test. In Barnard's Monte Carlo test, the observed value of the test statistic is compared with the values of the test statistic computed in each of m simulated samples of the same size as the given sample. For a fixed significance level, α, Barnard suggested fixing r so that $r/(m+1) \approx \alpha$, and then basing the test decision on whether fewer than r of the simulated values of the test statistics exceeded the observed value of the test statistic. Hope (1968) and Marriott (1979) studied the power of the test and found that the power of Monte Carlo tests can be quite good even for r as small as 5, which would require only a small number of simulated samples, often of the order of 100 (which would correspond to a test of level approximately 0.05, if $r = 5$).

Monte Carlo tests are used when the distribution of the test statistic is not known. To use a Monte Carlo test, however, the distribution of the random component in an assumed model (usually the "error" term in the model) must be known. In a Monte Carlo test, new, artificial samples are generated using the distribution of the random component in the model. For each Monte Carlo sample thus generated, the value of the test statistic is computed. As in randomization tests and bootstrap tests, the observed value of the test statistic is compared to the values computed from the artificial samples. The proportion of the simulated test statistics more extreme than the value of the test statistic computed from the given sample is determined and is taken as the "p-value" of the computed test statistic. The decision rule is then the same as in usual statistical hypothesis testing.

There are several applications of Monte Carlo tests that have been reported. Many involve spatial distributions of species of plants or animals. Manly (1991) gives several examples of Monte Carlo test in biology. Agresti (1992) describes Monte Carlo tests in contingency tables. Forster, McDonald, and Smith (1996) describe conditional Monte Carlo tests based on Gibbs sampling in log-linear and logistic models.

5.5.2   Bootstrap Methods

Resampling methods involve use of several samples taken from the one observed sample. The objective of the resampling methods may be to estimate the variance or the bias of an estimator. Of course, if we can estimate the bias, we may be able to correct for it, so as to have an unbiased estimator. Resampling can be used to estimate the significance level of a test statistic or to form confidence intervals for a parameter. The methods can be used when very little is known about the underlying distribution.

One form of resampling, the "jackknife" and its generalizations, involves forming subsamples without replacement from the observed sample. Although the jackknife has more general applications, one motivation for its use is that in certain situations it can reduce the bias of an estimator.

Another form of resampling, the nonparametric "bootstrap", involves forming samples, with replacement, from the observed sample. The basic idea of the nonparametric bootstrap is that, because the observed sample contains all the available information about the underlying population, the observed sample can be considered *to be* the population; hence, the distribution of any relevant test statistic can be simulated by taking random samples from the "population" consisting of the original sample.

Suppose a sample $x_1, x_2, \ldots, x_n$ is to be used to estimate a population parameter, θ. We form a statistic T that estimates θ. We wish to know the sampling distribution of T so as to set confidence intervals for our estimate of θ. The sampling distribution of T is often intractable in applications of interest.

The basic idea of the bootstrap is that the true population can be approximated by an infinite population in which each of the n sample points are equally likely. Using this approximation of the true population, we can approximate distributions of statistics formed from the sample. The basic tool is the empirical distribution function, which is the distribution function of the finite population that is to be used as an approximation of the underlying population of interest. We denote the empirical cumulative distribution function based on sample of size n as $P_n(\cdot)$. It is defined as

$$P_n(x) = \frac{1}{n} \sum_{i=1}^{n} I_{(-\infty, x]}(x_i),$$

where the indicator function $I_S(x)$ is defined by

$$I_S(x) \;=\; 1, \text{ if } x \in S;$$
$$=\; 0, \text{ otherwise.}$$

The parameter is a functional of a population distribution function:

$$\theta = \int g(x)\,\mathrm{d}P(x).$$

The estimator is often the same functional of the empirical distribution function:

$$T = \int g(x)\,\mathrm{d}P_n(x)$$

Various properties of the distribution of T can be estimated by use of "bootstrap samples", each of the form $\{x_1^*, x_2^*, \ldots, x_n^*\}$, where the x_i^*'s are chosen from the original x_i's with replacement.

The problem in its broadest setting is to find a functional f_t (from some class of functionals) that allows us to relate the distribution function of the sample P_n to the population distribution function P, that is, such that

$$\mathrm{E}(f_t(P, P_n) \mid P) = 0.$$

We then wish to estimate $h(\int g(x)\,\mathrm{d}P(x))$. The h presents special problems.

For example, suppose we wish to estimate

$$\theta = \left(\int x\,\mathrm{d}P(x)\right)^r.$$

The estimator

$$T = \left(\int x\,\mathrm{d}P_n(x)\right)^r = \bar{x}^r$$

is biased. Correcting for the bias is equivalent to finding t that solves the equation

$$f_t(P, P_n) = T(P_n) - \theta(P) + t$$

so that f_t has zero expectation with respect to P.

The *bootstrap principle* suggests repeating this whole process: We now take a sample from the "population" with distribution function P_n. We look for $f_t^{(1)}$ so that

$$\mathrm{E}\left(f_t^{(1)}(P_n, P_n^{(1)}) \mid P_n\right) = 0,$$

where $P_n^{(1)}$ is the empirical distribution function for a sample from the discrete distribution formed from the original sample.

The difference is we know more about this equation because we know more about P_n. Our knowledge of P_n comes either from parametric assumptions about the underlying distribution, or from just working with the discrete distribution in which the original sample is given equal probabilities at each mass point.

Nonparametric Bootstrap

In the nonparametric bootstrap, P_n is the empirical distribution function, and $P_n^{(1)}$ is the empirical distribution function of a sample drawn at random (with replacement) from the finite population with distribution function P_n.

Find f_t (i.e., t) so that

$$E(f_t(P, P_n) \mid P) = 0$$

or

$$E(T(P_n) - \theta(P) + t \mid P) = 0.$$

We can change the problem to the sample:

$$E(T(P_n^{(1)}) - T(P_n) + t_1 \mid P_n) = 0,$$

whose solution is

$$t_1 = T(P_n) - E(T(P_n^{(1)}) \mid P_n).$$

An estimate with less bias is therefore

$$T_1 = 2T(P_n) - E(T(P_n^{(1)}) \mid P_n).$$

We may be able to compute $E(T(F_2) \mid F_1)$, but generally we must resort to Monte Carlo methods to estimate it. The Monte Carlo estimate is based on m random samples each of size n, taken with replacement from the original sample.

The basic nonparametric bootstrap procedure is to take m random samples each of size n and *with replacement* from the given set of data, the original sample $x_1, x_2, \ldots, x_n$; and for each sample, compute an estimate T_j of the same functional form as the original estimator T. The distribution of the T_j's is related to the distribution of T. The variability of T about θ can be assessed by the variability of T_j about T; the bias of T can be assessed by the mean of $T_j - T$.

Parametric Bootstrap

In the parametric bootstrap, the distribution function of the population of interest, F, is assumed known up to a finite set of unknown parameters, λ. The estimate of P, $\widehat{P}$, instead of P_n, is P with λ replaced by its sample estimates (of some kind). Likewise $\widehat{P}^{(1)}$ is formed from $\widehat{P}$ by using estimates that are the same function of a sample from a population with distribution function $\widehat{P}$. This population is more like the original assumed population, and the sample is not just drawn with replacement from the original sample, as is done in the nonparametric bootstrap.

In a parametric bootstrap procedure, the first step is to obtain estimates of the parameters that characterize the distribution within the assumed family. After this, the procedure is very similar to that described above: generate m

random samples each of size n from the estimated distribution, and for each sample, compute an estimate T_j of the same functional form as the original estimator T. The sampling is done with replacement, as usual in random number generation. The distribution of the T_j's is used to make inferences about the distribution of T.

Extensive general discussions of the bootstrap are available in Efron and Tibshirani (1993), Shao and Tu (1995), and Davison and Hinkley (1997).

5.6 Evaluating a Posterior Distribution

In the Bayesian approach to data analysis, the parameter in the probability model for the data is taken to be a random variable. In this approach, the model is a prior distribution of the parameter together with a conditional distribution of the data given the parameter. Instead of statistical inference about the parameters being based on the estimation of the distribution of a statistic, Bayesian inference is based on the estimation of the conditional distribution of the parameter given the data. This conditional distribution of the parameter is called the posterior. Rather than the data being reduced to a statistic whose distribution is of interest, the data are used to form the conditional distribution of the parameter.

The prior distribution of the parameter, of course, has a parameter, and so the Bayesian analysis must take into account this "hyperparameter". (When we speak of a "parameter", we refer to an object that may be a scalar, a vector, or even some infinite-dimensional object such as a function. Generally, however, "parameter" refers to a real scalar or vector.) The hyperparameter for the parameter leads to a hierarchical model with the phenomenon of interest, modeled by the "data" random variable, having a probability distribution conditional on a parameter, which has a probability distribution conditional on another parameter, and so on. With an arbitrary starting point for the model, we may have the following random variables:

hyperprior parameter: Ψ

    prior parameter: Φ

        data model parameter: Θ

           data: Y

The probability distributions of these random variables (starting with Φ) are:

prior parameter, ϕ, with hyperprior distribution $p_\Phi(\cdot)$

    data model parameter, θ, with prior distribution $p_{\Theta|\phi}(\cdot)$

        data, y, with distribution $p_{Y|\theta,\phi}(\cdot)$

The distributions of interest are the conditional distributions given the data, the posterior distributions. These densities are obtained by forming the joint distributions using the hierarchical conditionals and marginals above and forming the marginals by integration. Except for specially chosen prior distributions, however, this integration is generally difficult. Moreover, in many interesting models, the integrals are multidimensional; hence, Monte Carlo is the tool of choice. The Markov chain Monte Carlo methods discussed in Sections 2.7 and 2.10 are used to generate the variates.

Instead of directly performing an integration, the analysis is performed by generating a sample of the parameter of interest, given the data. This sample is used for inference about the posterior distribution (see Gelfand and Smith, 1990, who used a Gibbs sampler).

Geweke (1991b) discusses general Monte Carlo methods for evaluating the integrals that arise in Bayesian analysis. Carlin and Louis (1996) and Gelman et al. (1995) provide extensive discussions of the methods of the data analysis and the associated computations. Also, Smith and Roberts (1993) give an overview of the applications of Gibbs sampling and other MCMC methods in Bayesian computations.

The computer program BUGS ("**B**ayesian inference **U**sing **G**ibbs **S**ampling") allows the user to specify a hierarchical and then evaluates the posterior by use of Gibbs sampling. See Gilks, Thomas, and Spiegelhalter (1992) or Thomas, Spiegelhalter, and Gilks (1992). The program is available from the authors at the Medical Research Council Biostatistics Unit at the University of Cambridge. Information can be obtained at the URL:

http://www.mrc-bsu.cam.ac.uk/bugs/

Exercises

5.1. Use Monte Carlo to estimate the expected value of the 5th order statistic from a sample of size 25 from a $N(0, 1)$ distribution. As with any estimate, you should also compute an estimate of the variance of your estimate. Compare your estimate of the expected value with the true expected value of 0.90501. Is your estimate reasonable? (Your answer must take into account your estimate of the variance.)

5.2. The "hit-or-miss" method is another Monte Carlo method to evaluate an integral. To simplify the exposition, let us assume $f(x) \geq 0$ on $[a, b]$, and we wish to evaluate the integral, $I = \int_a^b f(x)\, dx$. First determine c such that $c \geq f(x)$ on $[a, b]$. Generate a random sample n of pairs of uniform deviates (x_i, y_i), in which x_i is from a uniform distribution over $[a, b]$, y_i is from a uniform distribution over $[0, c]$, and x_i and y_i are independent. Let m be the number of pairs such that $y_i \leq f(x_i)$. Estimate I by $c(b-a)m/n$. (Sketch a picture, and you can see why it is called hit-or-miss. Notice also the similarity of the hit-or-miss method to acceptance/rejection. It can be generalized by allowing the y's to arise from a more general distribution, i.e., a majorizing density. Another way to think of the hit-or-miss method is as importance sampling in which the sampling of the weighting function is accomplished by acceptance/rejection.)

(a) Is this a better method than the crude method described in Section 5.1? (What does this question mean? Think "bias" and "variance". To simplify the computations for answering this question, consider the special case in which $[a, b]$ is $[0, 1]$ and $c = 1$. For further discussion, see Hammersley and Handscomb, 1964.)

(b) Suppose f is a probability density and the hit-or-miss method is used. Consider the set of the m x_i's for which $y_i \leq f(x_i)$. What can you say about this set with respect to the probability density f?

Because a hit-or-miss estimate is a rational fraction, the methods are subject to granularity. See the discussion following the Buffon's needle problem in Exercise 5.4 below.

5.3. Obtain a Monte Carlo estimate of the base of the natural logarithm, e. Your estimate should include confidence bounds.

5.4. The French naturalist Comté de Buffon showed that the probability that a needle of length l thrown randomly onto a grid of parallel lines with distance d $(> l)$ apart intersects a line is

$$\frac{2l}{\pi d}.$$

(a) Write a program to simulate the throwing of a needle onto a grid of parallel lines and obtain an estimate of π. Remember all estimates are to be accompanied by an estimate of the variability, that is, a standard deviation, a confidence interval, etc.

(b) Derive the variance of your estimator in Exercise 5.4a and determine the optimum ratio for l and r.

(c) Now consider a variation. Instead of a grid of just parallel lines, use a rectangular grid, as in Figure 5.1. For simplicity, let us estimate

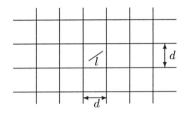

Figure 5.1: Buffon's Needle on a Square Grid

$\theta = 1/\pi$. If n_v is the number of crossings of vertical lines, and n_h is the number of crossings of horizontal lines, then for a total of n tosses, both

$$\frac{dn_v}{2ln} \quad \text{and} \quad \frac{dn_h}{2ln}$$

are unbiased estimates of θ. Show that the estimator

$$\frac{dn_v + dn_h}{4ln}$$

is more than twice as efficient (has variance less than half) as either of the separate estimators. This is an application of antithetic variates.

(d) Perlman and Wichura (1975) consider this problem and show that a complete sufficient statistic for θ is $n_1 + n_2$, where n_1 is the number of times the needle crosses exactly one line and n_2 is the number of times the needle crosses two lines. Derive an unbiased estimator of θ based on this sufficient statistic and determine its variance. Determine the optimum ratio for l and r.

(e) Now consider another variation: use a triangular grid, as in Figure 5.2. Write a program to estimate π using the square grid and

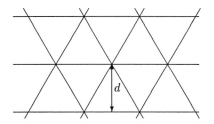

Figure 5.2: Buffon's Needle on a Triangular Grid

using the triangular grid. Using 10,000 trials, obtain estimates of π using each type of grid. Obtain an estimate of the variance of each of your estimators.

(C. R. Rao in *Statistics and Truth*, Council of Scientific & Industrial Research, New Delhi, 1989, relates some interesting facts about reports of Buffon needle experiments. An Italian mathematician, Lazzarini, reported in 1901 that he had obtained an estimate of 3.1415929 for π, using the simple Buffon method of one set of parallel lines. This estimate differs from the true value only in the seventh decimal place. This estimate came from 1808 successes in 3408 trials. Why 3408, instead of 3400 or 3500? The answer may be related to the fact that the best rational approximation of π for integers less than 15,000 is $\frac{355}{113}$, and that is the value Lazzarini reported, to seven places. The next rational approximation that is as close is $\frac{52163}{16604}$. In Lazzarini's experiment, $\frac{l}{d} = \frac{5}{6}$, so he could have obtained his accurate approximation only if the number of trials he used was $213, 426, \ldots, 3408, 3621, \ldots$, or else much larger numbers. The choice of 3408 does appear suspect. One wonders if exactly 1808 successes occurred the first time he tried 3408 trials.)

5.5. Consider the following Monte Carlo method to evaluate the integral

$$I = \int_a^b f(x) \, dx.$$

Generate a random sample of uniform order statistics $x_{(1)}, x_{(2)}, \ldots, x_{(n)}$ on the interval (a, b), and define $x_{(0)} = a$ and $x_{(n+1)} = b$. Estimate I by $\widehat{I}$:

$$\frac{1}{2} \left(\sum_{i=1}^n (x_{(i+1)} - x_{(i-1)}) f(x_{(i)}) + (x_{(2)} - a) f(x_{(1)}) + (b - x_{(n-1)}) f(x_{(n)}) \right)$$

Determine the variance of $\widehat{I}$. How would this method compare with crude Monte Carlo in efficiency?

5.6. Obtain a simplified expression for the variance of the Monte Carlo estimator (5.5) on page 133.

5.7. Consider the integral

$$\int_0^\infty x^2 \sin(\pi x) e^{-\frac{x}{2}} \, dx.$$

(a) Use crude Monte Carlo with an exponential weight to estimate the integral.

(b) Use antithetic exponential variates to estimate the integral.

(c) Compare the variance of the estimator in Exercise 5.7a with that of the estimator in Exercise 5.7b by performing the estimation 100 times in each case with sample sizes of 1000.

5.8. Generate a pseudorandom sample of size 100 from a $N(0, 1)$ distribution that has a sample mean of 0 and sample variance of 1.

5.9. Consider two estimators for the mean of a t distribution with 2 degrees of freedom. (The mean obviously is 0, but we may think of this problem as arising in a model that has an unknown mean plus a stochastic component that has a heavy-tailed distribution much like a t with 2 degrees of freedom.) Two unbiased estimators are T, the α-trimmed mean (which is the mean of the remaining data after the most extreme α fraction of the data has been discarded), and U, the α-Winsorized mean (which is the mean of the dataset formed by shrinking the most extreme α fraction of the data to the extreme order statistics of the central $1 - \alpha$ fraction of the data).

(a) Estimate the difference in the variance of T and U for samples of size 100 from a t distribution with 2 degrees of freedom.

(b) Now estimate the difference in the variance of $\widehat{E}(T^2 - U^2)$ and the variance of $\widehat{V}(T) - \widehat{V}(U)$, where the estimators $\widehat{E}$ and $\widehat{V}$ are based on Monte Carlo samples of size 500. (Read this problem carefully! See equation (5.8).)

5.10. Consider a common application in statistics: three different treatments are to be compared by applying them to randomly selected experimental units. This, of course, usually leads us to "analysis of variance", using a model such as $y_{ij} = \mu + \alpha_i + e_{ij}$, with the standard meanings of these symbols, and the usual assumptions about the random component e_{ij} in the model. Suppose, instead of the usual assumptions, we assume the e_{ij} have independent and identical beta distributions centered on zero (the density is proportional to $(\frac{c+x}{2c})^p(\frac{c-x}{2c})^p$ over $(-c,c)$). Describe how you would perform a Monte Carlo test instead of the usual AOV test. Be clear in stating the alternative hypothesis.

5.11. Let $S = \{x_1, x_2, \ldots, x_n\}$ be a random sample from a population with mean μ, variance σ^2, and distribution function P. Let $\widehat{P}$ be the empirical distribution function. Let $\bar{x}$ be the sample mean for S. Let $S_b = \{x_1^*, x_2^*, \ldots, x_n^*\}$ be a random sample taken with replacement from S. Let $\bar{x}_b^*$ be the sample mean for S_b.

(a) Show that
$$E_{\widehat{P}}(\bar{x}_b^*) = \bar{x}.$$

(b) Show that
$$E_P(\bar{x}_b^*) = \mu.$$

(c) Note that in the questions above, there was no replication of the bootstrap sampling. Now suppose we take B samples S_b, compute $\bar{x}_b^*$ for each, and compute
$$V = \frac{1}{B-1}\sum_b (\bar{x}_b^* - \overline{\bar{x}_b^*})^2.$$

Derive $E_{\widehat{P}}(V)$.

(d) Derive $E_P(V)$.

Chapter 6

Quality of Random Number Generators

Ziff (1992) describes a simulation requiring a total of 6×10^{12} random numbers and using a few months computing time on about a dozen workstations running simultaneously. Research work like this that depends so heavily on random numbers emphasizes the need for high-quality random number generators. Yet, as we have seen, not all random number generators are good ones (Park and Miller, 1988, and Ripley, 1988).

Only random number generators that have solid theoretical properties and that have also successfully passed a battery of tests should be used. Even so, often in Monte Carlo applications it is appropriate to construct ad hoc tests that are sensitive to departures from distributional properties that are important in the given application. For example, in using Monte Carlo methods to evaluate a one-dimensional integral, autocorrelations of order one may not be harmful, but they may be disastrous in evaluating a two-dimensional integral. If the routines for generating random deviates from nonuniform distributions use exact methods, their quality depends almost solely on the quality of the underlying uniform generator. Nevertheless, it is often advisable to employ an ad hoc goodness-of-fit test for the transformations that are to be applied.

6.1 Analysis of the Algorithm

For a generator that yields m different values, k-tuples of successive numbers in the output would ideally fall on m^k points uniformly spaced through a k-dimensional hypercube. As we have seen, however, k-tuples of the output from pseudorandom number generators lie on somewhat restrictive lattices in a k-dimensional hypercube.

In a simple linear congruential generator with period m, no point can repeat within a single period; therefore, there can be only m points in the lattice generated by such a generator. Although, as we have seen, there is no reason to use only simple linear congruential generators, they are widely used, and so assessing their lattice structure is of interest. The lattice structure of these common types of congruential generators can be assessed by means of two analytic "tests" of the generator.

The *spectral test* determines the maximum distance between adjacent parallel hyperplanes defined by the lattice. As we saw heuristically in Section 1.1, the smaller the maximum distance is, the better the output of the generator covers the hypercube.

The spectral test for a linear congruential generator does not use the output of the generator, but rather computes values analogous to a discrete Fourier transform of the output using the recurrence formula of the congruential relationship itself. For a linear congruential generator with multiplier a and modulus m, let $d_k(m, a)$ represent the maximum distance between adjacent parallel hyperplanes. Now, for any k-dimensional lattice with m points, there is a lower bound, $d_k^*(m)$, on $d_k(m, a)$ (see Knuth, 1981); hence, a useful measure for a given multiplier is the ratio

$$S_k(m, a) = \frac{d_k^*(m)}{d_k(m, a)}. \tag{6.1}$$

The closer this ratio is to 1, the better the generator is with respect to this measure. Coveyou and MacPherson (1967) describe a spectral test method that computes the denominator in equation (6.1). Their method was improved by Dieter (1975) and Knuth (1981), and is described in detail on pages 98 through 100 of the latter reference. Hopkins (1983) gives a computer program to implement the Coveyou/MacPherson test.

L'Ecuyer (1988) computes values of $S_k(2^{31} - 1, a)$ for some specific values of k and a. For example, for the multiplier 16 807 (see Section 1.1),

k	2	3	4	5	6
$S_k(2^{31} - 1,\ 16\,807)$	.34	.44	.58	.74	.65

and for the multiplier 950 706 376 (see Exercise 1.7, page 39),

k	2	3	4	5	6
$S_k(2^{31} - 1,\ 950\,706\,376)$	.86	.90	.87	.83	.83

(Although often one sees values of $S_k(m, a)$ given to several digits of precision, maybe because the computational algorithms used are probably accurate to 6 or 7 digits, anything beyond 1 or 2 digits is not very meaningful.)

Another assessment of the badness of the lattice is the *lattice test* of Marsaglia (1972a), which is ratio of the longest edge of the smallest parallelepiped in k dimensions to that of the shortest edge. Atkinson (1980) implemented

Marsaglia's lattice test and illustrated its use on some generators. Eichenauer and Niederreiter (1988) give examples that show that the lattice test is quite weak.

Other ways of examining the structure are to perform statistical analyses of the output of the generator. Neither the spectral test nor the lattice test use statistical analysis.

The general problem of long-range correlations in sequences from multiplicative congruential generators has been studied by Eichenauer-Herrmann and Grothe (1989) and by De Matteis and Pagnutti (1990). In general, for most generators the correlations die out very early, that is, at very low lags; and then the correlations grow again at large lags. At a lag equal to the period, the generator becomes 1, of course. For some generators, the correlation may become relatively large at some intermediate lags. In most cases, the correlations are not large enough to be statistically significant if the generated sequence would be assumed to arise from a random (instead of pseudorandom) process. The fact that the correlations are not statistically significant does not mean that these systematic correlations would not be so bad as to invalidate a Monte Carlo study.

Beyer, Roof, and Williamson (1971) and Beyer (1972) describe a measure of the uniformity of the lattice of the output of a random number generator in terms of the ratio of the shortest vector to the longest vector in the Minkowski reduced basis of the lattice. A basis for a lattice in $\mathbb{R}^d$ is a linearly-independent generating set of the lattice. Given a set of linearly independent vectors $\{v_1, v_2, \ldots, v_d\}$ in $\mathbb{R}^d$, a lattice is the set of vectors w of the form $\sum_{i=1}^{d} z_i v_i$, where z_i are integers. (See Figure 1.1, page 9.) The set of vectors $\{v_i\}$ is a basis for the lattice. The basis is a Minkowski reduced basis if

$$\|v_k\| \le \left\| \sum_{i=1}^{d} z_i v_i \right\|, \quad \text{for } 1 \le k \le d,$$

for all sets of d integers z_i such that the greatest common divisor of the set $\{z_k, z_{k+1}, \ldots, z_d\}$ is 1. (The symbol $\|v\|$ denotes the Euclidean norm of the vector v.) The ratio of the length of the shortest vector to the longest length,

$$\frac{\|v_d\|}{\|v_1\|},$$

is called the *Beyer ratio*. A large Beyer ratio (close to 1) indicates good uniformity. It is not easy to compute the Beyer ratio for a given generator. Afflerbach and Grothe (1985) give an algorithm for computing the ratio for linear congruential generators. They used the algorithm up to dimension of 20. Eichenauer-Herrmann and Grothe (1990) give upper bounds for the ratio for linear congruential generators that are applicable to higher dimensions. The problem of determining the shortest vectors defining a lattice has been shown to be NP-complete.

The *discrepancy* of a set of points $P = \{p_1, p_2, \ldots, p_n\}$ in the unit hypercube is

$$D_n(P) = \sup_J |F_n(J) - V(J)|, \qquad (6.2)$$

where the supremum is taken over all subintervals J of the unit hypercube, $F_n(J)$ is the number of points in $P \cap J$ divided by n, and $V(J)$ is the volume of J. The discrepancy is a measure of the uniformity of spread of a given set of points within another set of points.

For a given generator of points, we may consider the discrepancy of arbitrary sets of points within an s-dimensional unit hypercube. In that case we use the notation $D^{(s)}$.

6.2   Empirical Assessments

In addition to analyzing the algorithm itself, we may evaluate a specific generator by analyzing the output stream it produces. This approach has the additional advantage of assessing the program that is actually used, so the question of whether the program correctly implements the algorithm does not arise. There are basically two general kinds of properties of a generator to assess: the elements in the *set* of deviates without regard to the order in which they were generated, that is, the "static" properties; and the patterns in the sequences of the deviates, that is, the "dynamic" properties.

In Section 1.1, page 6, we discussed some ways of assessing the output of a random number generator. Those tests addressed the structure of the output and how well the output covered the space. Other kinds of tests are statistical tests applied to a sample of the output. An important source of information about the quality of a generator is a specific simulation. Whatever is known about the physics of the simulation can be compared with the results of the simulation. Both statistical tests of the output stream and the anecdotal evidence of a specific simulation are called empirical tests.

6.2.1   Statistical Tests

The static statistical tests are goodness-of-fit tests, such as chi-squared tests or Kolmogorov-Smirnov tests, for various combinations and transformations of the numbers in the output stream. These tests are of infinite variety. An easy test is just a goodness-of-fit test for uniformity in various dimensions. The dynamic statistical tests are for the independence of the output. These tests address patterns in the output stream; an example is a test for serial correlation of various lags. Kennedy and Gentle (1980) describe a variety of statistical tests of both types.

Fishman and Moore (1982, 1986) developed an extensive suite of goodness-of-fit tests based on various transformations of the sample. They applied the tests to linear congruential generators with the same modulus but different multipliers, in order to identify good multipliers.

All it takes to devise an empirical test is a little imagination. Many tests are based on familiar everyday random processes, such as occur in gaming. A "poker test", for example, is a chi-squared goodness-of-fit test for simulated poker hands. The idea is to devise a scheme to use a random number generator for simulating deals from a poker deck, and then to compare the sampled frequency of various poker hands with the expected frequency of such hands.

Marsaglia (1985, 1995) describes a battery of goodness-of-fit tests called "DIEHARD". Source code for these tests is available in a CD-ROM (Marsaglia, 1995). Most of these tests are performed on integers in the interval $(0, 2^{31} - 1)$ that are hypothesized to be realizations of a discrete uniform distribution with mass points being the integers in that range. These tests indicate the kinds of possibilities:

- birthday spacings test
  For this test, choose m birthdays in a year of n days. List the spacings between the birthdays. If j is the number of values that occur more than once in that list, then j has an asymptotic Poisson distribution with mean $m^3/(4n)$. Various groups of bits in the binary representation of the hypothesized uniform integers are used to simulate birthday spacings; goodness-of-fit tests are performed, yielding p-values; other groups of bits are used and tests performed to yield more p-values; and so on. Then a goodness-of-fit test is performed on the p-values.

- overlapping 5-permutation test
  Each set of five consecutive integers can be in one of 120 states, for the 5! possible orderings of five numbers. The test is on the observed numbers of states and transition frequencies.

- binary rank test for 31×31 matrices
  The leftmost 31 bits of 31 random integers from the test sequence are used to form a 31×31 binary matrix over the field $\mathbb{G}(2)$. The rank of the matrix is determined. Each time this is done, counts are accumulated for matrices with ranks of 31, of 30, of 29 and of 28 or less. A chi-squared test is performed for these four outcomes.

- binary rank test for 32×32 matrices
  This is similar to the test above.

- binary rank test for 6×8 matrices
  This is similar to the tests above, except that the rows of the matrix are specified bytes in the integer words.

- bitstream test
  Using the stream of bits from the random number generator, form 20-bit words, beginning with each successive bit, that is, the words overlap. The bitstream test compares the observed number of missing 20-bit words in a string of 2^{21} overlapping 20-bit words with the approximate distribution of that number.

- overlapping-pairs-sparse-occupancy test
  For this test 2-letter "words" are formed from an alphabet of 1024 letters. The letters in a word are determined by a specified ten bits from a 32-bit integer in the sequence to be tested, and the bits defining the letters overlap. The test counts the number of missing words, that is, combinations that do not appear in the entire sequence being tested. The count has an asymptotic normal distribution, and that is the basis for the goodness-of-fit test, when many sequences are used.

- overlapping-quadruples-sparse-occupancy
  This is similar to the test above. The null distribution of the test statistic is very complicated; interestingly, a parameter of the null distribution of the test statistic was estimated by Monte Carlo.

- DNA test
  This is similar to the tests above, except that it uses 10-letter words built on a 4-letter alphabet (the DNA alphabet).

- count-the-1's test on a stream of bytes
  This test is based on the binomial (8, 0.5) distribution of 1's in uniform random bytes. Rather than testing this directly, however, counts of 1's are grouped into 5 sets $\{0, 1, 2\}$, $\{3\}$, $\{4\}$, $\{5\}$, and $\{6, 7, 8\}$; that is, if a byte has no 1's, exactly one 1, or exactly two 1's, it is counted in the first group. Next, overlapping sequences of length 5 are formed and the counts of each of the 5^5 combinations are obtained. A chi-squared test is performed on the counts. The test takes into account the covariances of the counts, and so is asymptotically correct. If the groups of counts of 1's are thought of as 5 letters, and the groups of bytes thought of as 5-letter words, the output of the sequence under test is similar to the typing of a "monkey at a typewriter hitting five keys" (with rather strange probabilities), so the test is sometimes called the "monkey at a typewriter test".

- count-the-1's test for specific bytes
  This is similar to the test above.

- parking lot test
  This test is based on the results of randomly packing circles of radius 1 about a center located randomly within a square of side 100. The test is based on the number of "cars parked" (i.e., non-overlapping circles packed) after a large number of attempts. The distribution of the test statistic, and hence its critical values, are determined by simulation.

- minimum distance test
  This test is based on the minimum distance between a large number of random points in a square. If the points are independent uniform, then the square of the minimum distance should be approximately exponentially distributed with mean dependent on the length of the side of the square

and the number of points. A chi-squared goodness-of-fit test is performed
on the p-values of a large number of tests for exponentiality.

- 3-D spheres test
  In this test, 4000 points are chosen randomly in a cube of edge 1000. At
  each point, a sphere is centered large enough to reach the next closest
  point. The volume of the smallest such sphere is approximately exponentially distributed with mean $\frac{120\pi}{3}$. Thus the radius cubed is exponential
  with mean 30. (The mean was obtained by extensive simulation.) A chi-
  squared goodness-of-fit test is performed on the p-values of a large number
  of tests for exponentiality.

- squeeze test
  This test is performed on floating-point numbers hypothesized to be from
  a $U(0, 1)$ distribution. Starting with $k = 2^{31}$, the test finds j, the number
  of iterations necessary to reduce k to 1, using the reduction $k = \lceil k * U \rceil$,
  with U from the stream being tested. Such j's are found 100,000 times,
  then counts for the number of times j was $\leq 6, 7, \ldots, 47, \geq 48$ are used to
  provide a chi-squared test.

- overlapping sums test
  This test is performed on floating-point numbers hypothesized to be from
  a $U(0, 1)$ distribution. Sums of overlapping subsequences of 100 uniforms
  are tested for multivariate normality by a chi-squared test several times
  and then a chi-squared goodness-of-fit test is performed on the p-values
  of the chi-squared tests.

- craps test
  This test simulates games of craps and counts the number of wins and
  the number of throws necessary to end each game, and then performs
  chi-squared goodness-of-fit tests on the observed counts.

- runs test
  See the following discussion.

One of the most effective of the dynamic tests is the runs test. A runs test
uses the number and/or length of successive runs in which the values of the
sequence are either increasing or decreasing. There are different ways these can
be counted. To illustrate one way of counting runs, consider the sequence:

$$1, 5, 4, 1, 3, 1, 3, 4, 7$$

There are 5 different runs in this sequence:

$$\text{up}: 1, 5; \quad \text{down}: 5, 4, 1; \quad \text{up}: 1, 3; \quad \text{down}: 3, 1; \quad \text{up}: 1, 3, 4, 7$$

The number of runs could be used as a test statistic, but a more powerful test
can be constructed using the lengths of the runs up and/or the runs down. The

lengths of the runs up as shown above are 2, 2, and 4. With small integers such as in this example, an issue may be how to handle ties, that is, two adjacent numbers that are equal. This issue is not relevant when the test is applied to the output of a generator that simulates a continuous uniform distribution.

A more common way of counting the lengths of the runs up (and analogously, the runs down) is to identify breakpoints where the sequence no longer continues increasing (or decreasing). The lengths of runs are then the lengths of the subsequences between breakpoints. For the sequence above we have breakpoints as indicated below:

$$1, 5 \quad | \quad 4 \quad | \quad 1, 3 \quad | \quad 1, 3, 4, 7$$

This gives 1 run-up of length 1; 2 runs-up of length 2; and 1 run-up of length 4. The runs test (more properly, the runs-up test and the runs-down test) is based on the covariance matrix of the random number of runs of each length. Usually only runs up to some fixed length, such as 7, are considered. (Runs greater than that are lumped together; the covariance matrix can be adjusted to fix the test.) The test statistic is the quadratic form in the counts with the covariance matrix, which has an asymptotic chi-squared distribution with degrees of freedom equal to the number of different lengths of runs counted. Grafton (1981) gives a Fortran program to compute the chi-squared test statistic for the runs test (with runs of length 1 through 6) and the significance level of the test statistic. The program is available from `statlib` as the *Applied Statistics* algorithm AS 157.

For a given test, the significance level of the test statistic, under the null hypothesis is $U(0, 1)$ (recall our discussion of the inverse CDF technique). This fact gives rise to a second-order test: perform a selected goodness-of-fit test many times and then perform a goodness-of-fit test on the p-values. This fact gives rise to a third-order test, and so on, ad infinitum.

Because the empirical tests are statistical tests, the ordinary principles of hypothesis testing apply. The results of a battery of tests must be interpreted carefully. As Marsaglia (1995) states (sic):

> By all means, do not, as a Statistician might, think that a $p < .025$ or $p > .975$ means that the RNG has "failed the test at the .05 level". Such p's happen among the hundreds that DIEHARD produces, even with good RNG's. So keep in mind that "p happens".

Fortunately, a "Statistician" understands statistical hypothesis testing better than some other testers of random number generators.

6.2.2   Anecdotal Evidence

The null hypothesis in the statistical tests of random number generators is that "the generator is performing properly"; therefore, failure to reject is not confirmation that the generator is a good one. Tests reported in the literature have often been inconclusive.

Another way that random number generators are tested is in their usage in simulations in which some of the output can be compared with what established theory would suggest. We call this anecdotal evidence. The first indication that the RANDU generator had problems came from anecdotal evidence in a simulation whose results did not correspond to theory (see Coldwell, 1974).

Ferrenberg, Landau, and Wong (1992) use some of the generators that meet the Park and Miller (1988) minimal standard (see page 18) to perform several simulation studies in which the correct answer was known. Their simulation results suggest that even some of the "good" generators could not be relied on in some simulations. In complex studies it is difficult to trace unexpected results to errors in the computer code or to some previously unknown phenomenon. Because of sampling error in simulations, there is always a question of whether results are due to a sample that is unusual. Vattulainen, Ala-Nissila, and Kankaala (1994) studied the methods of Ferrenberg, Landau, and Wong, and determined that the anomalous results were indeed due to defects in the random number generator. Vattulainen, Ala-Nissila, and Kankaala conducted further studies on these generators as well as generators of other types and found that their simulations often did not correspond to reality. Selke, Talapov, Shchur (1993), however, found that the Park-Miller minimum standard generator performed much better than the R250 generator in a specific simulation of particle behavior. These kinds of inconclusive results lead us to repeat the advice given above to employ an ad hoc goodness-of-fit test for the specific application.

Cuccaro, Mascagni, and Pryor (1994) describe some approaches for testing parallel random number generators. A major problem in random number generation in parallel is the inability to synchronize the computations. This is easy to appreciate even in a simple case of acceptance/rejection; if one processor must do more rejections than another, that processor will get behind the other one.

6.3 Quasirandom Numbers

The idea of stratification to reduce sampling variance, as we discussed in Section 5.3.3, can be extended; as the number of strata increases, the sample size in each stratum decreases. In the limit there is no randomness remaining in the sampling.

In our discussion of the Monte Carlo estimator in equation (5.2) (page 132), we indicated that the expressions for the expectation and variance were for samples from a random variable. For a deterministic sequence, instead of the expectation of the estimator, we might consider its limit; and so we might take into account the discrepancy (equation (6.2), page 154).

For pseudorandom numbers generated by any of the methods we have discussed, the limit is taken over a cyclic finite set, and the supremum in the discrepancy is a maximum. Suppose, instead of the pseudorandom numbers re-

sulting from the generators we have discussed, we explicitly address the problem
of discrepancy. Then, instead of being concerned with $E(\hat{\theta})$, we might consider

$$\lim_{n\to\infty} \hat{\theta} = \lim_{n\to\infty} (b-a) \frac{\sum f(y_i)}{n}.$$

A moment's pause, of course, tells us this kind of limit does not apply to the
real world of computing, with its finite set of numbers. Nevertheless, it is useful
to consider a deterministic sequence with low discrepancy. The objective is that
any finite subsequence fill the space uniformly.

Several such sequences have been proposed, such as van der Corput sequences, Halton sequences (Halton, 1960), Faure sequences, Sobol' sequences
(Sobol', 1967, 1976), and Niederreiter sequences (Niederreiter, 1988). These
sequences are called *quasirandom* sequences. Whereas pseudorandom sequences
or pseudorandom generators attempt to simulate randomness, quasirandom sequences are decidedly *not* random. The objective for a (finite) pseudorandom
sequence is for it to "look like" a sequence of realizations of i.i.d. uniform random variables; but for a (finite) quasirandom sequence the objective is that it
fill a unit hypercube as uniformly as possible.

Quasirandom sequences correspond to samples from a $U(0,1)$ distribution.
(Contrast this statement with the statement that "pseudorandom sequences
simulate random samples from a $U(0,1)$ distribution".) The techniques of
Chapters 2 and 3 can therefore be used to generate quasirandom sequences
that correspond to samples from nonuniform distributions. For the methods
that yield one nonuniform deviate from each uniform deviate, such as the inverse CDF method, everything is straightforward. For other methods that use
multiple independent uniforms for each nonuniform deviate, the quasirandom
sequence may be inappropriate. The quasirandom method does not simulate
independence.

6.3.1   Halton Sequences

Halton sequences are generalizations of van der Corput sequences. A Halton
sequence is formed by reversing the digits in the representation of some sequence
of integers in a given base. Although this can be done somewhat arbitrarily,
a straightforward way of forming a d-dimensional Halton sequence $x_1, x_2, \ldots,$
where $x_i = (x_{i1}, x_{i2}, \ldots, x_{id})$ is first to choose d bases, $b_1, b_2, \ldots, b_d$, perhaps
the first d primes. The jth base will be used to form the jth component of each
vector in the sequence. Then begin with some integer m and

1. choosing t_{mj} suitably large, represent m in each base:

$$m = \sum_{k=0}^{t_{mj}} a_{mk} b_j^k, \quad j = 1, \ldots d,$$

2. form

$$x_{ij} = \sum_{k=0}^{t_{mj}} a_{mk} b_j^{k-t_{mj}-1}, \quad j = 1, \ldots d,$$

3. set $m = m + 1$ and repeat.

Suppose, for example, $d = 3$, $m = 15$, and we use the bases 2, 3, and 5. We form $15 = 1111_2$, $15 = 120_3$, and $15 = 30_5$, and deliver the first x as $(0.1111_2, 0.021_3, 0.03_5)$, or $(0.937500, 0.259259, 0.120000)$.

The Halton sequences are acceptably uniform for lower dimensions, up to about 10. For higher dimensions, however, the quality of the Halton sequences degrades rapidly because the two-dimensional planes occur in cycles with decreasing periods. Generalized Halton sequences have been proposed and studied by Braaten and Weller (1979), Hellekalek (1984), Faure (1986), and Krommer and Ueberhuber (1994). The basic idea is to permute the a_{mk}'s in step 2. The Faure sequences are formed in this way.

Kocis and Whiten (1997) suggest a "leaped Halton sequence". In this method the cycles of the Halton sequence are destroyed by using only every l^{th} Halton number, where l is a prime different from all of the bases $b_1, b_2, \ldots, b_d$.

6.3.2 Sobol' Sequences

A Sobol' sequence is based on a set of "direction numbers", $\{v_i\}$. The v_i are

$$v_i = \frac{m_i}{2^i},$$

where the m_i are odd positive integers less than 2^i; and the v_i are chosen so that they satisfy a recurrence relation using the coefficients of a primitive polynomial in the Galois field $\mathbb{G}(2)$,

$$f(z) = z^p + c_1 z^{p-1} + \cdots + c_{p-1} z + c_p$$

(compare equation (1.12), page 27). For $i > p$, the recurrence relation is

$$v_i = c_1 v_{i-1} \oplus c_2 v_{i-2} \oplus \cdots \oplus c_p v_{i-p} \oplus \lfloor v_{i-p}/2^p \rfloor,$$

where $\oplus$ denotes bitwise binary exclusive-or. An equivalent recurrence for the m_i is

$$m_i = 2c_1 m_{i-1} \oplus 2^2 c_2 m_{i-2} \oplus \cdots \oplus 2^p c_p m_{i-p} \oplus m_{i-p}.$$

As an example, consider the primitive polynomial (1.16) from page 27,

$$x^4 + x + 1.$$

The corresponding recurrence is

$$m_i = 8m_{i-3} \oplus 16m_{i-4} \oplus m_{i-4}.$$

If we start with $m_1 = 1$, $m_2 = 1$, $m_3 = 3$, and $m_4 = 13$, for example, we get

$$
\begin{aligned}
m_5 &= 8 \oplus 16 \oplus 1 \\
&= 01000(\text{binary}) \oplus 10000(\text{binary}) \oplus 00001(\text{binary}) \\
&= 11001(\text{binary}) \\
&= 25.
\end{aligned}
$$

The ith number in the Sobol' sequence is now formed as

$$
x_i = b_1 v_1 \oplus b_2 v_2 \oplus b_3 v_3 \oplus \cdots,
$$

where $\cdots b_3 b_2 b_1$ is the binary representation of i.

Antonov and Saleev (1979) show that equivalently the Sobol' sequence can be formed as

$$
x_i = g_1 v_1 \oplus g_2 v_2 \oplus g_3 v_3 \oplus \cdots, \tag{6.3}
$$

where $\cdots g_3 g_2 g_1$ is the binary representation of a particular Gray code evaluated at i. (A *Gray code* is a function, $G(i)$, on the nonnegative integers such that the binary representations of $G(i)$ and $G(i+1)$ differ in exactly one bit.) The binary representation of the Gray code used by Antonov and Saleev is

$$
\cdots g_3 g_2 g_1 = \cdots b_3 b_2 b_1 \oplus \cdots b_4 b_3 b_2.
$$

(This is the most commonly used Gray code, which yields function values $0, 1, 3, 2, 6, 7, 5, 4, \ldots$) The Sobol' sequence from (6.3) can be generated recursively by

$$
x_i = x_{i-1} \oplus v_r,
$$

where r is determined so that b_r is the rightmost zero bit in the binary representation of $i - 1$.

Bratley and Fox (1988) discuss criteria for starting values, $m_1, m_2, \ldots$. (The starting values used in the example with the primitive polynomial above satisfy those criteria.)

6.3.3   Comparisons

Empirical comparisons of various quasirandom sequences that have been reported in the literature are somewhat inconclusive. Sarkar and Prasad (1987) compare the performance of pseudorandom and quasirandom sequences in the solution of integral equations by Monte Carlo methods and find no difference in the performance of the two quasirandom sequences they studied: the Halton and Faure sequences. Fox (1986), on the other hand, finds the performance of Faure sequences to be better than that of Halton sequences. This is confirmed by Bratley and Fox (1988), who also find that the performance of the Sobol' sequence is roughly the same as that of the Faure sequence. The empirical results reported in Bratley, Fox, and Niederreiter (1992) show inconclusive differences between Sobol' sequences and Niederreiter sequences. Tezuka (1993) describes an analog of Halton sequences and a generalization of Niederreiter sequences, and then shows that these extensions are related.

Although in some cases it is simple to make comparisons between the performance of pseudorandom and quasirandom sequences, there is a fundamental difference in the nature of the error bounds appropriate for Monte Carlo methods and for other numerical algorithms. Hickernell (1995) compares a certain type of error bounds for quadrature using pseudorandom and quasirandom sequences, and shows that the quasirandom sequences resulted in smaller bounds for errors. Bouleau and Lépingle (1994), quoting Pagès and Xiao, give comparative sample sizes required by pseudorandom and Halton and Faure sequences to achieve the same precision for quadrature in various dimensions from 2 to 20. Precision, in this case, is approximated using the asymptotic convergence of the quadrature formulas.

6.3.4 Variations

Quasirandom sequences that cover domains other than hypercubes have also been studied. Beck and Chen (1987) review work on low-discrepancy sequences over various types of regions.

Fang and Wang (1994), defined a type of quasirandom sequence of higher dimensions that they called NT-nets ("number theory" nets). Fang and Li (1997) gave a method for generating random orthogonal matrices based on an NT-net.

Combination generators (see Section 1.7, page 32) for quasirandom sequences can also be generated. It generally would not be useful to combine another sequence with a quasirandom sequence, except to shuffle the quasirandom sequence. The shuffling generator could be either a quasirandom generator or a pseudorandom generator, but a pseudorandom generator would be more likely to be effective. When a combination generator is composed of one or more quasirandom generators and one or more pseudorandom generators, it is called a hybrid generator. Braaten and Weller (1979) describe a hybrid method in which a pseudorandom generator is used to scramble a quasirandom sequence.

6.3.5 Some Examples of Applications

There are several applications of Monte Carlo reported in the literature that use quasirandom numbers. For example, Shaw (1988) uses quasirandom sequences instead of the usual pseudorandom sequences for evaluating integrals arising in Bayesian inference. Do (1991) uses quasirandom sequences in a Monte Carlo bootstrap. Joy, Boyle, and Tan (1996) empirically comparing the use of Faure sequences with pseudorandom sequences in valuing financial options of various types, find that the quasirandom sequences had better convergence rates in that application. Also in an application in fincancial derivatives, Papageorgiou and Traub (1996) compared the performance of a generalized Faure sequence with a Sobol' sequence. They concluded that the Faure sequence was superior in that problem.

6.3.6 Computations

Halton sequences are easy to generate (Exercise 6.4a asks you to write a program to do so). Fox (1986) gives a program for Faure sequences; Bratley and Fox (1988) give a program for Sobol' sequences (using Gray codes, as mentioned above); and Bratley, Fox, and Niederreiter (1994) give a program for Niederreiter sequences.

6.4 Programming Issues

In addition to the quality of the algorithm for generation of random numbers, there are also many issues relating to the quality of the computer implementation of the random number generator. Because of the limitations of representing numbers exactly, a computer program rarely corresponds exactly to a mathematical expression. Sometimes a poor program completely negates the positive qualities of the algorithm it was intended to implement. The programming considerations relevant to implementing a random number generator are often subtle and somewhat different from those arising in other mathematical software (see Gentle, 1990).

To the extent that widely-used and well-tested software for random number generation is available, it should be used instead of software developed ad hoc. Software for random number generation is discussed in Chapter 7.

Exercises

6.1. Devise a "dice test" (see Marsaglia's "craps test" in Section 6.2). Your test should accept the output of a uniform random number generator (you can assume either discrete or continuous) and simulate the roll of a pair of dice to obtain a total (i.e., the pairs (1,3), (2,2), (2,2), and (3,1) are not distinguished). For a specified number of rolls, your test should be a chi-squared test with 11 degrees of freedom (meaning that the number of rolls should be greater than some fixed number).

Implement your test in the language of your choice. Then apply it to some random number generators, for example, the generator you wrote in Exercise 1.2 of Chapter 1, and whatever system random number is available to you, such as in S-Plus or the IMSL Libraries.

Now put your dice test in the inner loop of a test program so as to perform the test many times, each time retaining the p-value. At the end of the inner loop do a chi-squared test on the p-values. Does this testing of the output of tests make sense? What about testing the output of tests of the output of tests?

6.2. Sullivan (1993) suggested the following test of a random number generator: For each of the seeds $1, 2, \ldots, N$, generate a sequence of length n.

Let I represent the index of the maximum element in each sequence. If n is considerably less than the period of the generator, we would expect I to be uniformly distributed over the integers $1, 2, \ldots, n$.

Using $N = 100$ and $n = 10000$, perform this test on the "minimal standard" generator you wrote in Exercise 1.6. What are your results? What is wrong with this test?

6.3. Test the Fortran or C RANDU generator you wrote in Exercise 1.4 of Chapter 1. Devise any statistical tests you want. (See Section 6.2, and use your imagination.) Does your generator pass all your tests? (Do not answer "no" until you have found a realistic statistical test that RANDU actually fails.)

6.4. Halton sequences.

(a) Write a subprogram to generate d-dimensional Halton sequences, using the integers j, $j+1$, $\ldots$, $j+n-1$, and using the first d primes as the bases for each succeeding component of the d-tuples. Your subprogram should accept as input j, d, n, and possibly the declared first dimension of the output array. Your subprogram should output an $n \times d$ array.

(b) Use the subprogram from Exercise 6.4a to evaluate the integral

$$\int_0^1 \int_0^2 y \sin(\pi x) \, dy \, dx.$$

(c) Apply the dice test you constructed in Exercise 6.1 to the Halton sequences.

6.5. Design and develop a hybrid combination generator that uses a pseudo-random number generator, perhaps such as you wrote in Exercise 1.7 of Chapter 1 (page 39), to shuffle the output of a quasirandom number generator, perhaps the Halton sequence generator in Exercise 6.4a. Be sure to provide appropriate user control to set the initial state of the combination generator.

Chapter 7

Software for Random Number Generation

Random number generators are widely available in a variety of software packages. As Park and Miller (1988) state, however, "good ones are hard to find".

Some programming languages such as C and Fortran 90 provide built-in random number generators. The revision of Ada currently called 9X also contains a built-in random number generator. In C the generator is the function `rand()` in `stdlib.h`. This function returns an integer in the range 0 through `RAND_MAX`, so the result must be normalized to the range $(0, 1)$. (The scaling should be done with care. Recall that it is desirable to have uniforms in $(0, 1)$ rather than $[0, 1]$; see Exercise 1.10, page 40.) The seed for the C random number generator is set in `srand()`.

In Fortran 90 the generator is the subroutine `random_number`, which returns $U(0, 1)$ numbers. (The user must be careful, however; the generator may yield either a 0 or a 1.) The seed can be set in the subroutine `random_seed`. The design of the Fortran 90 module as a subroutine yields a major advantage over the C function in terms of efficiency. (Of course, because Fortran 90 has the basic advantage of arrays, the module could have been designed as an array function and would still have had an advantage over the C function.)

A basic problem with either the built-in generator of C or of Fortran 90 is the problem of portability discussed in Section 1.6, page 31. The bindings are portable, but neither generator will necessarily generate the same sequence on different platforms.

Given a uniform random number generator, it is usually not too difficult to generate variates from other distributions using the techniques discussed in Chapters 2 and 3. For example, in Fortran 90, the inverse CDF technique for generating a random deviate from a Bernoulli distribution with parameter π shown in Algorithm 2.3, page 47, can be implemented by the code in Figure 7.1.

Implementing one of the simple methods to convert a uniform deviate to that of another distribution may not be as efficient as a special method for

```
integer, parameter      :: n  = 100   ! INITIALIZE THIS
real, parameter (pi)    :: pi = .5    ! INITIALIZE THIS
real, dimension (n)     :: uniform
real, dimension (n)     :: bernoulli
call random_number (uniform)
where (uniform .le. pi)
      bernoulli = 1.0
elsewhere
      bernoulli = 0.0
endwhere
```

Figure 7.1: A Fortran 90 Code Fragment to Generate n Bernoulli Random Deviates with Parameter π

the target distribution; and, as we have indicated, those special methods may be somewhat complicated. The IMSL Libraries and S-Plus have a number of modules that use efficient methods to generate variates from several of the more common distributions. Matlab has a basic uniform generator, rand, and a standard normal generator, randn.

A number of Fortran or C programs are available in collections published by *Applied Statistics* and by *ACM Transactions on Mathematical Software*. These collections are available online at statlib and netlib, respectively. See the bibliography for more information.

The Guide to Available Mathematical Software, or GAMS (see the bibliography) can be used to locate special software for various distributions.

7.1   The User Interface for Random Number Generators

Software for random number generation must provide a certain amount of control by the user, including the ability to

- set or retrieve the seed

- select seeds that yield separate streams

- possibly select the method from a limited number of choices.

Whenever the user invokes a random number generator for the first time in a program or a session, the software should not require the specification of a seed, but it should allow the user to set it if desired. If the user does not specify the seed, the software should use some mechanism, such as accessing the system clock, to form a "random" seed. On a subsequent invocation of the random number generator, unless the user specifies a seed, the software should use the last value of the seed from the previous invocation. This means that the routine for generating random numbers must produce a "side effect"; that is, it changes something other than the main result. It is a basic tenet of software engineering that careful note must be taken of side effects. At one time side effects were generally to be avoided. In object-oriented programming, however, objects may encapsulate many entities, and as the object is acted upon, any of the components may change; so in object-oriented software side effects are to be expected. In object-oriented software for random number generation, the state of the generator is an object.

Another issue to consider in the design of a user interface for a random number generator is whether the output is a single value (and an updated seed) or an array of values. Although a function that produces a single value as the C function **rand()** is convenient to use, it can carry quite a penalty in execution time because of the multiple invocations required to generate an array of random numbers. It is generally better to provide both single- and multivalued procedures for random number generation, especially for the basic uniform generator.

7.2   Controlling the Seeds in Monte Carlo Studies

There are three reasons the user must be able to control the seeds in Monte Carlo studies: for testing of the program, for use of blocks in Monte Carlo experiments, and for combining results of Monte Carlo studies.

In the early phases of programming for a Monte Carlo study it is very important to be able to test the output of the program. To do this it is necessary to use the same seed from one run of the program to another.

Controlling seeds in a parallel random number generator is much more complicated than in a serial generator. Performing Monte Carlo computations in parallel requires some way of insuring the independence of the parallel streams (see Section 1.8, page 36).

7.3   Random Number Generation in IMSL Libraries

For doing Monte Carlo studies, it is usually better to use a software system with a compilable programming language, such as Fortran or C. Not only do such systems provide more flexibility and control, but the programs built in the compiler languages execute faster. To do much work in such a system, however, a library or routines both to perform the numerical computations in the inner loop of the Monte Carlo study and to generate the random numbers driving the study are needed.

The IMSL Libraries contain a large number of routines for random number generation. The libraries are available in both Fortran and C, each providing the same capabilities and with essentially the same interface within the two languages. In Fortran the basic uniform generator is provided in both function and subroutine forms. The uniform generator allows the user to choose among seven different algorithms: a linear congruential generator with modulus of $2^{31} - 1$ and with three choices of multiplier, each with or without shuffling; and the generalized feedback shift generator described by Fushimi (1990), which has a period of $2^{521} - 1$. The multipliers that the user can choose are the "minimal standard" one of Park and Miller (1988), which goes back to Lewis, Goodman, and Miller (1969) (see page 18) and two of the "best" multipliers found by Fishman and Moore (1982, 1986), one of which is used in Exercise 1.7, of Chapter 1.

User control is provided through a number of utility routines that allow choice of the basic uniform generator. For whatever choice is in effect, that form of the uniform generator will be used for whatever type of pseudorandom events are to be generated. The states of the generators are maintained in a `common` block (for the simple congruential generators, the state is a single seed; for the shuffled generators and the GFSR generator, the state is maintained in a table). There are utility routines for setting and saving states of the generators and a utility routine for obtaining a seed to skip ahead a fixed amount.

There are routines to generate deviates from most of the common distributions. Most of the routines are subroutines but some are functions. The algorithms used often depend on the values of the parameters, so as to achieve greater efficiency. The routines are available in both single and double precision. (Double precision is more for the purpose of convenience for the user than it is for increasing accuracy of the algorithm.)

A single precision IMSL Fortran subroutine for generating from a specific distribution has the form

> **rn***name* (*number, parameter_1, parameter_2, ..., output_array*)

where "*name*" is an identifier for the distribution, "*number*" is the number of random deviates to be generated, "*parameter_i*" are parameters of the distribution, and "*output_array*" is the output argument with the generated deviates. The Fortran subroutines generate variates from standard distributions, so location and scale parameters are not included in the argument list. The subroutine and formal arguments to generate gamma random deviates, for example, are

> **rngam (nr, a, r)**

where **a** is the shape parameter (α) of the gamma distribution. The other parameter in the common two-parameter gamma distribution (usually called β) is a scale parameter. The deviates produced by the routine **rngam** have a

scale parameter of 1; hence, for a scale parameter of **b**, the user would follow the call above with a call to a BLAS routine:

```
sscal (nr, b, r, 1)
```

For general distributions, there are routines for an alias method and for table lookup, for either discrete or continuous distributions. The user specifies a discrete distribution by providing a vector of the probabilities at the mass points, and specifies a continuous distribution by giving the values of the cumulative distribution function at a chosen set of points. In the case of a discrete distribution, the generation can be done either by an alias method or by an efficient table lookup method. For a continuous distribution, a cubic spline is first fit to the given values of the cumulative distribution function, and then an inverse CDF method is used to generate the random numbers from the target distribution. Another routine uses the Thompson-Taylor data-based scheme to generate deviates from an unknown population from which only a sample is available.

Other routines in the IMSL Libraries generate various kinds of time series, random permutations, and random samples. The routine **rnuno**, which generates order statistics from a uniform distribution, can be used to generate order statistics from other distributions.

All of the IMSL routines for random number generation are available in both Fortran and C. The C functions have more descriptive names, such as **random_normal**. Also, the C functions may allow specification of additional arguments, such as location and scale parameters. For example, **random_normal** has optional arguments **IMSLS_MEAN** and **IMSLS_VARIANCE**.

Controlling the State of the Generators

Figure 7.2 illustrates the way to save the state of an IMSL generator and then restart it. The functions to save and to set the seed are **rnget** and **rnset**.

```
    call rnget (iseed)      ! save it
    call rnun (nr, y)       ! get sample, analyze, etc.
...
    call rnset (iseed)      ! restore seed
    call rnun (nr, yagain)  ! will be the same as y
```

Figure 7.2: Fortran Code Fragment to Save and Restart a Random Sequence Using the IMSL Library

In a library of numerical routines such as the IMSL Libraries, it is likely that some of the routines will use random numbers in regular deterministic computations, such as an optimization routine generating random starting points. In a well-designed system, before a routine in the system uses a random number generator in the system, it will retrieve the current value of the seed if one has been set, use the generator, and then reset the seed to the former value. This allows the user to control the seeds in the routines called directly.

7.4   Random Number Generation in S-Plus

Random number generators in S-Plus are all based upon a single uniform random number generator that is a combination of a linear congruential generator and a Tausworthe generator.

In S-Plus there are some basic functions with one of the forms

> r*name* (*number* [, *parameters*])

or

> r*name* (*number, parameters* [, *parameters*])

where "*name*" is an identifier for the distribution, "*number*" is the number of random deviates to be generated, "*parameters*" are parameters of the distribution, and

> [·]

indicates optional information. For example,

> rnorm (100)

generates 100 normal random numbers with the default values of the parameters (zero mean and unit variance), and

> rgamma (100,5)

generates 100 random numbers from a gamma distribution with shape parameter of 5. (The shape parameter is required.)

Identifiers of distributions include the following.

unif	uniform	norm	normal
t	*t*	chisq	chi-squared
f	*F*	lnorm	lognormal
exp	exponential	gamma	gamma
weibull	Weibull	cauchy	Cauchy
beta	beta	logis	logistic
stab	stable	binom	binomial
nbinom	negative binomial	geom	geometric
hyper	hypergeometric	pois	Poisson
wilcox	Wilcoxon rank sum statistic		

The number of deviates to be generated is specified as the first argument of the function, as either the value of the scalar, truncated to an integer if necessary, or the length of the object.

For distributions with standard forms, such as the normal, the parameters may be optional, in which case they take on default values if they are not specified. For other distributions, such as the *t*, there are required parameters. Optional parameters are both positional and keyword.

For example, the normal variate generation function is

```
rnorm (n, mean=0, sd=1)
```

so,

`rnorm (n)`	yields n normal $(0,1)$ variates
`rnorm (n, 100, 10)`	yields n normal $(100,100)$ variates
`rnorm (n, 100)`	yields n normal $(100,1)$ variates
`rnorm (n, sd=10)`	yields n normal $(0,100)$ variates

Note that S-Plus considers one of the natural parameters of the normal distribution to be the standard deviation (the scale), rather than the variance, as is more common.

The function `sample` generates a random sample with or without replacement. Sampling with replacement is equivalent to generating random numbers from a (finite) discrete distribution. The mass points and probabilities can be specified in optional arguments.

```
uu <- runif(n, 0, 2)
eu <- 2*sum(log(uu+1)*uu^2*(2-uu)^3)/n
```

Order statistics in S-Plus can be generated using the beta distribution and the inverse distribution function. For example, 10 maximum order statistics from normal samples of size 30 can be generated by

```
x <- qnorm(rbeta(10,30,1))
```

Controlling the State of the Generators

S-Plus uses an object called `.Random.seed` to maintain the state of the random number generators. Anytime random number generation is performed in S-Plus, if `.Random.seed` does not exist in the user's working directory, it is created. If it exists, it is used to initiate the pseudorandom sequence, and then is updated after the sequence is generated. Setting a different working directory will change the state of the random number generator.

The function `set.seed(i)` sets the value of the `.Random.seed` object in the working directory. The argument `i` is an integer between 0 and 1000, and each value represents a state of the generator, which is "far away" from the other states that can be set in `set.seed`.

To save the state of the generator, just copy `.Random.seed` into a named object; and to restore, just copy the named object back into `.Random.seed`, as in Figure 7.3.

Functions in S-Plus that use the random number generators have the side effect of changing the state of the generators, so the user must be careful in Monte Carlo studies where the computational nuclei, such as `ltsreg` for robust regression, for example, invoke an S-Plus random number generator. In this case, the user must retrieve the state of the generator prior to calling the function, and then reset the state prior to the next invocation of a random number generator.

```
oldseed <- .Random.seed   # save it
y <- rnunif(1000)          # get sample, analyze, etc.
...
.Random.seed <- oldseed    # restore seed
yagain <- rnorm(1000)      # will be the same as y
```

Figure 7.3: Code Fragment to Save and Restart a Random Sequence Using S-Plus

Monte Carlo in S-Plus

Explicit loops in S-Plus execute very slowly. For that reason, it is best to use array arguments to functions in S-Plus, rather than to loop over scalar values of the arguments. Consider, for example, the problem of evaluating the integral

$$\int_0^2 \log(x+1)x^2(2-x)^3\,\mathrm{d}x.$$

This could be estimated in a loop as follows:

```
# First, initialize n.
uu <- runif(n, 0, 2)
eu <- 0
for (i in 1:n) eu <- eu + log(uu[i]+1)*uu[i]^2*(2-uu[i])^3
eu <- 2*eu/n
```

A much more efficient way, without the **for** loop, but still using the uniform, is

```
uu <- runif(n, 0, 2)
eu <- 2*sum(log(uu+1)*uu^2*(2-uu)^3)/n
```

Alternatively, using the beta density as a weight function, we have

```
eb <- (64/35)*sum(log(2*rbeta(n,3,4)+1))/n
```

Of course, if we recognize the relationship of the integral to the beta distribution, we would not use Monte Carlo as the method of integration.

Exercises

7.1. Identify as many random number generators as you can that are available on your computer system. Try to determine what method each uses. Do the generators appear to be of high quality?

7.2. Consider the problem of evaluating the integral

$$\int_{-\pi}^{\pi}\int_0^4\int_0^\infty x^2y^3\sin(z)(\pi+z)^2(\pi-z)^3e^{-\frac{x}{2}}\,\mathrm{d}x\,\mathrm{d}y\,\mathrm{d}z.$$

Note the gamma and beta weighting functions.

(a) Write a Fortran or C program to use the IMSL Libraries to evaluate this integral by Monte Carlo. Use a sample of size 1000, and save the state of the generator, so you can restart it. Now, use a sample of size 10,000, starting where you left off in the first 1000. Combine your two estimates.

(b) Now, do the same thing in S-Plus.

(c) Now, do the same thing in Fortran 90, using its built-in random number functions. You may use other software to evaluate special functions if you wish.

7.3. Obtain the programs for Algorithm 738 for generating quasirandom numbers (Bratley, Fox, and Niederreiter, 1994) from the *Collected Algorithms of the ACM*. The programs are in Fortran and may require a small number of system-dependent modifications, which are described in the documentation embedded in the source code.

Devise some tests for Monte Carlo evaluation of multidimensional integrals and compare the performance of Algorithm 738 with that of a pseudorandom number generator. (Just use any convenient pseudorandom generator available to you.) The subroutine TESTF accompanying Algorithm 738 can be used for this purpose.

Can you notice any difference in the performance?

Chapter 8

Monte Carlo Studies in Statistics

In statistical inference, certain properties of the test statistic or estimator must be assumed to be known. In simple cases, under rigorous assumptions we have complete knowledge of the statistic. In testing a mean of a normal distribution, we use a t statistic, and we know its exact distribution. In other cases, however, we may have a perfectly reasonable test statistic, but know very little about its distribution. For example, suppose a statistic T, computed from a differenced time series, could be used to test the hypothesis that the order of differencing is sufficient to yield a series with a zero mean. If the standard deviation of T is known under the null hypothesis, that value may be used to construct a test that the differencing is adequate. This, in fact, was what Erastus Lyman de Forest did in the 1870s, in one of the earliest documented Monte Carlo studies of a statistical procedure. De Forest studied ways of smoothing a time series by simulating the data using cards drawn from a box. A description of De Forest's Monte Carlo study is given in Stigler (1978). Stigler (1991) also describes other Monte Carlo simulation by nineteenth-century scientists, and suggests that "Simulation, in the modern sense of that term, may be the oldest of the stochastic arts".

Another early use of Monte Carlo was the sampling experiment (using biometric data recorded on pieces of cardboard) that led W. S. Gosset to the discovery of the distribution of the t-statistic and the correlation coefficient (see Student, 1908a, 1908b).

Major advances in Monte Carlo techniques were made during World War II and after by mathematicians and scientists working on problems in atomic physics. (In fact, it was the group led by John von Neumann and S. M. Ulam who coined the term "Monte Carlo" to refer to these methods.) The use of Monte Carlo by statisticians gradually increased from the time of De Forest; but after the widespread availability of digital computers, the usage greatly expanded.

In the mathematical sciences, including statistics, simulation has become an important tool in the development of theory and methods. For example, if the properties of an estimator are very difficult to work out analytically, a Monte Carlo study may be conducted to estimate those properties.

Often the Monte Carlo study is an informal investigation whose main purpose is to indicate promising research directions. If a "quick and dirty" Monte Carlo study indicates that some method of inference has good properties, it may be worth the time of the research worker in developing the method and perhaps doing the difficult analysis to confirm the results of the Monte Carlo study.

In addition to quick Monte Carlo studies that are mere precursors to analytic work, Monte Carlo studies often provide a significant amount of the available knowledge of the properties of statistical techniques, especially under various alternative models. Eddy and Gentle (1985) reported that in 1983 25% of the articles in one major statistical journal included Monte Carlo studies. In another major statistical journal that same year, 35% of the articles reported on Monte Carlo studies. In the 1995 volume of *Journal of the American Statistical Association*, 49% of the articles reported on Monte Carlo studies.

8.1   Simulation as an Experiment

A simulation study that incorporates a random component is an experiment. The principles of statistical design and analysis apply just as much to a Monte Carlo study as they do to any other scientific experiment. The Monte Carlo study should adhere to the same high standards of any scientific experimentation:

- control

- reproducibility

- efficiency

- careful and complete documentation

In simulation, *control*, among other things, relates to the fidelity of a *nonrandom* process to a *random* process. The experimental units are only simulated. Questions about the computer model must be addressed (tests of the random number generators, etc.)

Likewise, *reproducibility* is predicated on good random number generators (or else on equally bad ones!). Portability of the random number generators enhances reproducibility and in fact can allow *strict* reproducibility. Reproducible research also requires preservation and documentation of the computer programs that produced the results (see Buckheit and Donoho, 1995).

The principles of good statistical design can improve the efficiency. Use of good designs (fractional factorials, etc.) can allow efficient simultaneous exploration of several factors. Also, there are often many opportunities to reduce the variance (improve the efficiency). Hammersley and Hanscomb (1964, page 8) note,

> ... statisticians were insistent that other experimentalists should design experiments to be as little subject to unwanted error as possible, and had indeed given important and useful help to the experimentalist in this way; but in their own experiments they were singularly inefficient, nay negligent in this respect.

Many properties of statistical methods of inference are analytically intractable. Asymptotic results, which are often easy to work out, may imply excellent performance, such as consistency with a good rate of convergence, but the finite sample properties are ultimately what must be considered. Monte Carlo studies are a common tool for investigating the properties of a statistical method, as noted above. In the literature, the Monte Carlo studies are sometimes called "numerical results". Some numerical results are illustrated by just one randomly generated dataset; others are studied by averaging over thousands of randomly generated sets.

In a Monte Carlo study there are usually several different things ("treatments" or "factors") that we want to investigate. As in other kinds of experiments, a factorial design is usually more efficient. Each factor occurs at different "levels", and the set of all levels of all factors that are used in the study constitute the "design space". The measured responses are properties of the statistical methods, such as their sample means and variances.

The factors commonly studied in Monte Carlo experiments in statistics include the following.

- statistical method (estimator, test procedure, etc.)

- sample size

- The problem for which the statistical method is being applied, that is, the "true" model, which may be different from the one for which the method was developed. Factors relating to the type of problem may be:
    - distribution of the random component in the model (normality?)
    - correlation among observations (independence?)
    - homogeneity of the observations (outliers?, mixtures?)
    - structure of associated variables (leverage?)

The factor whose effect is of primary interest is the statistical method. The other factors are generally just blocking factors. There is, however, usually an interaction between the statistical method and these other factors.

As in physical experimentation, observational units are selected for each point in the design space and measured. The measurements, or "responses" made at the same design point, are used to assess the amount of random variation, or variation that is not accounted for by the factors being studied. A comparison of the variation among observational units at the same levels of all factors with the variation among observational units at different levels is the basis for a decision as to whether there are real (or "significant") differences at the different levels of the factors. This comparison is called analysis of variance. The same basic rationale for identifying differences is used in simulation experiments.

A fundamental (and difficult) question in experimental design is how many experimental units to observe at the various design points. Because the experimental units in Monte Carlo studies are generated on the computer, they are usually rather inexpensive. The subsequent processing (the application of the factors, in the terminology of an experiment) may be very extensive, however, so there is a need to design an efficient experiment.

8.2   Reporting Simulation Experiments

The reporting of a simulation experiment should receive the same care and consideration that would be accorded the reporting of other scientific experiments. Hoaglin and Andrews (1975) outline the items that should be included in a report of a simulation study. In addition to a careful general description of the experiment, the report should include mention of the random number generator used, any variance-reducing methods employed, and a justification of the simulation sample size. The *Journal of the American Statistical Association* includes these reporting standards in its style guide for authors.

Closely related to the choice of the sample size is the standard deviation of the estimates that result from the study. The sample standard deviations actually achieved should be included as part of the report. Standard deviations are often reported in parentheses beside the estimates with which they are associated. A formal analysis, of course, would use the sample variance of each estimate to assess the significance of the differences observed between points in the design space; that is, a formal analysis of the simulation experiment would be a standard analysis of variance.

The most common method of reporting the results is by means of tables, but a better understanding of the results can often be conveyed by graphs.

8.3   An Example

One area of statistics in which Monte Carlo studies have been used extensively is robust statistics. This is because the finite sampling distributions of many robust statistics are very difficult to work out, especially for the kinds of underlying distributions for which the statistics are to be studied. An example of an important study of robust statistics is described by Andrews et al. (1972), who introduced and examined many alternative estimators of location for samples from univariate distributions. This study, which involved many Monte Carlo experiments, described innovative methods of variance reduction and was very influential in subsequent Monte Carlo studies reported in the statistical literature.

As an example of a Monte Carlo study, we describe a simple experiment to assess the robustness of a statistical test in linear regression analysis. The purpose of this example is to illustrate some of the issues in designing a Monte Carlo experiment. The results of this small study are not of interest here. There are many important issues about the robustness of the procedures that we do not address in this example.

The Problem

Consider the simple linear regression model,

$$Y = \beta_0 + \beta_1 x + \mathrm{E},$$

where a response or "dependent variable", Y, is modeled as a linear function of a single regressor or "independent variable", x, plus a random variable, E, called the "error", hence, Y is also a random variable. The statistical problem is to make inferences about the unknown, constant parameters β_0 and β_1, and about distributional parameters of the random variable, E. The inferences are made based on a sample of n pairs (y_i, x_i), with which are associated unobservable realizations of the random error, ϵ_i, and which are assumed to have the relationship

$$y_i = \beta_0 + \beta_1 x_i + \epsilon_i. \tag{8.1}$$

We also generally assume that the realizations of the random error are independent and are unrelated to the value of x.

For this example, let us consider just the specific problem of testing the hypothesis

$$\mathrm{H}_0\colon \beta_1 = 0, \tag{8.2}$$

versus the universal alternative. If the distribution of E is normal and we make the additional assumptions above about the sample, the optimal test for the hypothesis (using the common definitions of optimality) is based on a least squares procedure that yields the statistic

$$t = \frac{\widehat{\beta}_1 \sqrt{(n-2)\sum(x_i - \bar{x})^2}}{\sqrt{\sum r_i^2}}, \tag{8.3}$$

where $\bar{x}$ is the mean of the x's, $\widehat{\beta}_1$ together with $\widehat{\beta}_0$ minimizes the function

$$L_2(b_0, b_1) = \sum_{i=1}^{n}(y_i - b_0 - b_1 x_i)^2,$$

and

$$r_i = y_i - (\widehat{\beta}_0 + \widehat{\beta}_1 x_i).$$

If the null hypothesis is true, t is a realization of a Student's t distribution with $n - 2$ degrees of freedom. The test is performed by comparing the p-value from the Student's t distribution with a preassigned significance level, α, or by comparing the observed value of t with a critical value. The test of the hypothesis depends on the estimates of β_0 and β_1 used in the test statistic t.

Often, a dataset contains outliers, that is, observations that have a realized error that is very large in absolute value, or observations for which the model is not appropriate. In such cases, the least squares procedure may not perform so well. We can see the effect of some outliers on the least squares estimates of β_0 and β_1 in Figure 8.1. For well-behaved data, as in the plot on the left, the least squares estimates seem to fit the data fairly well. For data with two outlying points, as in the plot on the right in Figure 8.1, the least squares estimates are affected so much by the two points in the upper left part of the graph that the estimates do not provide a good fit for the bulk of the data.

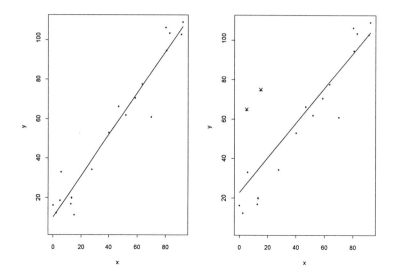

Figure 8.1: Least Squares Fit Using Two Datasets that are the Same Except for Two Outliers

Another method of fitting the linear regression line that is robust to outliers in E is to minimize the absolute values of the deviations. The least absolute values procedure chooses estimates of β_0 and β_1 so as to minimize the function

$$L_1(b_0, b_1) = \sum_{i=1}^{n} |y_i - b_0 - b_1 x_i|.$$

Figure 8.2 shows the same two data sets as before with the least squares (LS) fit and the least absolute values (LAV) fit plotted on both graphs. We see the least absolute values fit does not change because of the outliers.

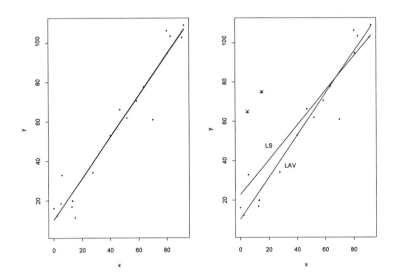

Figure 8.2: Least Squares Fits and Least Absolute Values Fits

Another concern in regression analysis is the unduly large influence that some individual observations exert on the aggregate statistics because the values of x in those observations lie at a large distance from the mean of all the x_i, that is, those observations whose values of the independent variables are outliers. The influence of an individual observation is called *leverage*. Figure 8.3 shows two datasets together with the least squares and the least absolute values fits for both. In both datasets, there is one value of x that lies far outside the range of the other values of x. All of the data in the plot on the left in Figure 8.3 lie relatively close to a line, and both fits are very similar. In the plot on the right, the observation with an extreme value of x also happens to have an outlying value of E. The effect on the least squares fit is marked, while the least absolute values fit is not affected as much. (Despite this example, least absolute values fits are generally not very robust to outliers at high leverage points; especially if there are multiple such outliers. There are other methods of fitting that are more robust to outliers at high leverage points. We refer the interested reader to Rousseeuw and Leroy, 1987, for discussion of these issues.)

Now we continue with our original objective in this example; that is, to evaluate ways of testing the hypothesis (8.2).

A test statistic analogous to the one in equation (8.3), but based on the

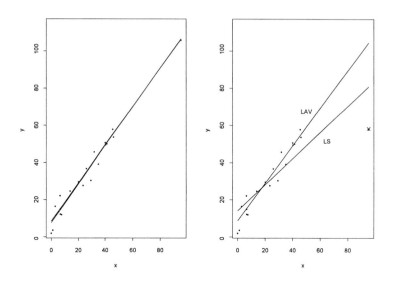

Figure 8.3: Least Squares and Least Absolute Values Fits

least absolute values fit, is

$$t_1 = \frac{2\tilde{\beta}_1 \sqrt{\sum (x_i - \bar{x})^2}}{(e_{(k_2)} - e_{(k_1)})\sqrt{n-2}}, \tag{8.4}$$

where $\tilde{\beta}_1$ together with $\tilde{\beta}_0$ minimizes the function

$$L_1(b_0, b_1) = \sum_{i=1}^{n} |y_i - b_0 - b_1 x_i|,$$

$e_{(k)}$ is the kth order statistic from

$$e_i = y_i - (\tilde{\beta}_0 + \tilde{\beta}_1 x_i),$$

k_1 is the integer closest to $(n-1)/2 - \sqrt{n-2}$, and k_2 is the integer closest to $(n-1)/2 + \sqrt{n-2}$. This statistic has an approximate Student's t distribution with $n-2$ degrees of freedom (see Birkes and Dodge, 1993, for example).

   If the distribution of the random error is normal, inference based on minimizing the sum of the absolute values is not nearly as efficient as inference based on least squares. This alternative to least squares should therefore be used with some discretion. Furthermore, there are other procedures that may warrant consideration. It is not our purpose here to explore these important issues in robust statistics, however.

The Design of the Experiment

At this point, we should have a clear picture of the problem: we wish to compare two ways of testing the hypothesis (8.2) under various scenarios. The data may have outliers, and there may be observations with large leverage. We expect that the optimal test procedure will depend on the presence of outliers, or more generally, on the distribution of the random error, and on the pattern of the values of the independent variable. The possibilities of interest for the distribution of the random error include

- the family of the distribution, that is, normal, double exponential, Cauchy, and so on

- whether the distribution is a mixture of more than one basic distribution, and if so, the proportions in the mixture

- the values of the parameters of the distribution, that is, the variance, the skewness, or any other parameters that may affect the power of the test.

In textbooks on design of experiments, the simple objective of an experiment is to perform a t test or an F test of whether different levels of response are associated with different treatments. Our objective in the experiment we are designing is to investigate and characterize the dependence of the performance of the hypothesis test on these factors. The principles of design are similar to those of other experiments, however.

It is possible that the optimal test of the hypothesis will depend on the the sample size or on the true values of the coefficients in the regression model, so some additional issues that are relevant to the performance of a statistical test of this hypothesis are the sample size and the true values of β_0 and β_1.

In the terminology of statistical models, the factors in our Monte Carlo experiment are the estimation method and the associated test, the distribution of the random error, the pattern of the independent variable, the sample size, and the true value of β_0 and β_1. The estimation method together with the associated test is the "treatment" of interest. The "effect" of interest, that is, the measured response, is the proportion of times that the null hypothesis is rejected using the two treatments.

We now can see our objective more clearly: for each setting of the distribution, pattern, and size factors, we wish to measure the power of the two tests. These factors are similar to blocking factors, except that there is likely to be an interaction between the treatment and these factors. Of course, the power depends on the nominal level of the test, α. It may be the case that the nominal level of the test affects the relative powers of the two tests.

We can think of the problem in the context of a binary response model,

$$\mathrm{E}(P_{ijklqsr}) = f(\tau_i, \delta_j, \phi_k, \nu_l, \alpha_q, \beta_{1s}), \tag{8.5}$$

where the parameters represent levels of the factors listed above (β_{1s} is the sth level of β_1), and $P_{ijklqsr}$ is a binary variable representing whether or not the test rejects the null hypothesis on the rth trial at the $(ijklqs)^{\text{th}}$ setting of the design factors. It is useful to write down a model like this to remind ourselves of the issues in designing an experiment.

At this point it is necessary to pay careful attention to our terminology. We are planning to use a statistical procedure (a Monte Carlo experiment) to evaluate a statistical procedure (a statistical test in a linear model). For the statistical procedure we will use, we have written a model (8.5) for the observations we will make. Those observations are indexed by r in that model. Let m be the sample size for each combination of factor settings. This is the Monte Carlo sample size. It is not to be confused with the data sample size, n, that is one of the factors in our study.

We now choose the levels of the factors in the Monte Carlo experiment.

- For the estimation method, we have decided on two methods: least squares and least absolute values. Its differential effect in the binary response model (8.5) is denoted by τ_i, for $i = 1, 2$.

- For the distribution of the random error, we choose three general ones:

  1. Normal $(0, 1)$

  2. Normal $(0, 1)$ with $c\%$ outliers from normal $(0, d^2)$

  3. Standard Cauchy

  We choose different values of c and d as appropriate. For this example, let us choose $c = 5$ and 20, and $d = 2$ and 5. Thus, in the binary response model (8.5), $j = 1, 2, 3, 4, 5, 6$.

- For the pattern of the independent variable, we choose three different arrangements:

  1. Uniform over the range:

  2. A group of extreme outliers:

  3. Two groups of outliers:

  In the binary response model (8.5), $k = 1, 2, 3$. We use fixed values of the independent variable.

- For the sample size, we choose three values: 20, 200, and 2,000. In the binary response model (8.5), $l = 1, 2, 3$.

- For the nominal level of the test, we choose two values: 0.01 and 0.05. In the binary response model (8.5), $q = 1, 2$.

- The true value of β_0 is probably not relevant, so we just choose $\beta_0 = 1$. We are interested in the power of the tests at different values of β_1. We expect the power function to be symmetric about $\beta_1 = 0$, and to approach 1 as $|\beta_1|$ increases.

The estimation method is the "treatment" of interest.

Restating our objective in terms of the notation introduced above, for each of two tests, we wish to estimate the power curve,

$$\Pr(\text{reject } H_0) = g(\beta_1 \mid \tau_i, \delta_j, \phi_k, \nu_l, \alpha_q),$$

for any combination $(\tau_i, \delta_j, \phi_k, \nu_l, \alpha_q)$. For either test, this curve should have the general appearance of the curve shown in Figure 8.4.

The minimum of the power curve should occur at $\beta_1 = 0$, and should be α. The curve should approach 1 symmetrically as $|\beta_1|$.

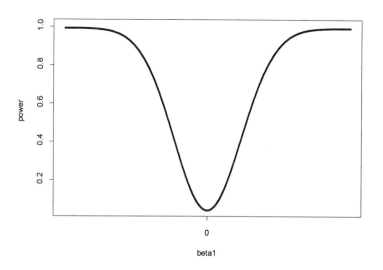

Figure 8.4: Power Curve for Testing $\beta_1 = 0$

To estimate the curve we use a discrete set of points; and because of symmetry, all values chosen for β_1 can be nonnegative. The first question is at what point does the curve flatten out just below 1. We might arbitrarily define the region of interest to be that in which the power is less than 0.99, approximately. The abscissa of this point is the maximum β_1 of interest. This point, say β_1^*, varies depending on all of the factors in the study. We could work this out in the least squares case for uncontaminated normal errors, using the noncentral Student's t distribution, but for all other cases, it is analytically intractable. Hence, we compute some preliminary Monte Carlo estimates to determine the maximum β_1 for each factor combination in the study.

To do a careful job of fitting a curve using a relatively small number of points, we would chose points where the second derivative is changing rapidly, and especially near points of inflection where the second derivative changes sign. Because the problem of determining these points for each combination of

(i, j, k, l, q) is not analytically tractable (otherwise we would not be doing the study!), we may conveniently chose a set of points equally spaced between 0 and β_1^*. Let us decide on five such points for this example. It is not important that the β_1^*'s be chosen with a great deat of care. The objective is that we be able to calcuate two power curves between 0 and β_1^* that are meaningful for comparisons.

The Experiment

The observational units in the experiment are the values of the test statistics (8.3) and (8.4). The measurements are the binary variables corresponding to rejection of the hypothesis (8.2). At each point in the factor space, there will be m such observations. If z is the number of rejections observed, the estimate of the power is z/m, and the variance of the estimator is $\pi(1 - \pi)/m$, where π is the true power at that point. (z is a realization of a binomial random variable with parameters m and π.) This leads us to a choice of the value of m. The coefficient of variation at any point is $\sqrt{(1 - \pi)/(m\pi)}$, which increases as π decreases. At $\pi = 0.50$, a 5% coefficient of variation can be achieved with a sample of size 400. This yields a standard deviation of 0.025. There may be some motivation to choose a slightly larger value of m, because the we can assume the minimum of π will be approximately the minimum of α. To achieve a 5% coefficient of variation at that point (i.e., at $\beta_1 = 0$) would require a sample of size approximately 160,000. That would correspond to a standard deviation of 0.0005, which is probably much smaller than we need. The sample size of 400 would yield a standard deviation of 0.005. Although that is large in a relative sense, it should be adequate for our purposes. Hence, we choose a Monte Carlo sample size of 400. We will, however, conduct the experiment in such a way that we can combine the results of this experiment with independent results from a subsequent experiment.

The experiment is conducted by running a computer program. The main computation in the program is to determine the values of the test statistics, and to compare them with their critical values so as to decide on the hypothesis. These computations need to be performed at each setting of the factors and for any given realization of the random sample.

We design a program that allows us to loop through the settings of the factors, and at each factor setting, to use a random sample. The result is a nest of loops. The program may be stopped and restarted, so we need to be able to control the seeds. (See Section 7.2, page 169.)

Recalling that the purpose of our experiment is to obtain estimates, we may now consider any appropriate methods of reducing the variance of those estimates. There is not much opportunity to apply the methods of variance reduction discussed in Section 5.3, but at least we might consider at what points to use common realizations of the pseudorandom variables. Because the things that we want to compare most directly are the powers of the tests, we perform the tests on the same pseudorandom datasets. Also, because we

are interested in the shape of the power curves we may want to use the same
pseudorandom datasets at each value of β_1, that is, to use the same set of
errors in the model (8.1). Finally, following similar reasoning, we may use
the same pseudorandom datasets at each value of pattern of the independent
variable. This implies that our program of nested loops has the structure shown
in Figure 8.5.

Initialize a table of counts.
      Fix the data sample size. (Loop over the sample sizes $n = 20$,
      $n = 200$, and $n = 2000$.)

            Generate a set of residuals for the linear regression
            model (8.1). (This is the loop of m Monte Carlo replications.)

                  Fix the pattern of the independent variable. (Loop over
                  patterns P_1, P_2, and P_3.)

                        Choose the distribution of the error term. (Loop
                        over the distributions D_1, D_2, D_3, D_4, D_5, and D_6.)

                              For each value of β_1, generate a set of obser-
                              vations (the y values) for the linear regression
                              model (8.1), and perform the tests using both
                              procedures and at both levels of significance.
                              Record results.
                        End distributions loop.
                  End patterns loop.
            End Monte Carlo loop.
      End sample size loop.
Perform computations of summary statistics.

Figure 8.5: Program Structure for the Monte Carlo Experiment

After writing a computer program with this structure, the first thing is to
test the program on a small set of problems and to determine appropriate values
of β_1^*. We should compare the results with known values at a few points. (As
mentioned earlier, the only points we can work out correspond to the normal
case with the ordinary t statistic. One of these points, at $\beta_1 = 0$, is easily
checked.) We can also check the internal consistency of the results. For example,
does the power curve increase? We must be careful, of course, in applying such
consistency checks, because we do not know the behavior of the tests in most
cases.

Reporting the Results

The report of this Monte Carlo study should address as completely as possible
the results of interest. The relative values of the power are the main points
of interest. The estimated power at $\beta_1 = 0$ is of interest. This is the actual

significance level of the test, and how it compares to the nominal level α is of particular interest.

The presentation should be in a form easily assimilated by the reader. This may mean graphs similar to Figure 8.4, except only the nonegative half, and with the tick marks on the horizontal axis. Two graphs, for the two test procedures, should be shown on the same set of axes. It is probably counterproductive to show a graph for each factor setting. (There are 108 combinations of factor settings.)

In addition to the graphs, tables may allow presentation of a large amount of information in a compact format.

The Monte Carlo study should be described so carefully that the study could be replicated exactly. This means specifying the factor settings, the loop nesting, the software and computer used, the seed used, and the Monte Carlo sample size. There should also be at least a simple statement explaining the choice the Monte Carlo sample size.

As mentioned earlier, the statistical literature is replete with reports of Monte Carlo studies. Some of these reports (and, likely, the studies themselves) are woefully deficient. An example of a careful Monte Carlo study and a good report of the study are given by Kleijnen (1977). He designed, performed, and reported a Monte Carlo study to investigate the robustness of a multiple ranking procedure. In addition to reporting on the study of the question at hand, another purpose of the paper was to illustrate the methods of a Monte Carlo study.

Exercises

8.1. Write a computer program to implement the Monte Carlo experiment described in Section 8.3. The S-Plus functions lsfit and l1fit or the IMSL Fortran subroutines rline and rlav can be used to calculate the fits. See Chapter 7 for discussions of other software you may use in the program.

8.2. Choose a recent issue of the *Journal of the American Statistical Association* and identify five articles that report on Monte Carlo studies of statistical methods. In each case, describe the Monte Carlo experiment.

   (a) What are the factors in the experiment?

   (b) What is the measured response?

   (c) What is the design space, that is, the set of factor settings?

   (d) What random number generators were used?

   (e) Critique the report in each article. Did the author(s) justify the sample size? Did the author(s) report variances or confidence intervals? Did the author(s) attempt to reduce the experimental variance?

8.3. Select an article you identified in Exercise 8.2 that concerns a statistical method that you understand and that interests you. Choose a design space that is not a subspace of that used in the article, but that has a nonnull intersection with it, and perform a similar experiment. Compare your results with those reported in the article.

Appendix A

Notation and Definitions

All notation used in this work is "standard". I have opted for simple notation, which, of course, results in a one-to-many map of notation to object classes. Within a given context, however, the overloaded notation is generally unambiguous. I have endeavored to use notation consistently.

This appendix is not intended to be a comprehensive listing of definitions. The Subject Index, beginning on page 243, is a more reliable set of pointers to definitions, except for symbols that are not words.

General Notation

Uppercase italic Latin and Greek letters, A, B, E, Λ, etc., are generally used to represent either matrices or random variables. Random variables are usually denoted by letters nearer the end of the Latin alphabet, X, Y, Z, and by the Greek letter E. Parameters in models, that is, unobservables in the models, whether or not they are considered to be random variables, are generally represented by lower case Greek letters. Uppercase Latin and Greek letters, especially P and Φ, are also used to represent cumulative distribution functions. Also, uppercase Latin letters are used to denote sets. Notice that uppercase Greek letters appear to be written in a roman font instead of an italic font.

Lowercase Latin and Greek letters are used to represent ordinary scalar or vector variables and functions. **No distinction in the notation is made between scalars and vectors**; thus, β may represent a vector and β_i may represent the ith element of the vector β. In another context, however, β may represent a scalar. All vectors are considered to be column vectors, although we may write a vector as $x = (x_1, x_2, \ldots, x_n)$. Transposition of a vector or a matrix is denoted by a superscript "T".

Subscripts generally represent indexes to a larger structure, for example, x_{ij} may represent the $(i,j)^{\text{th}}$ element of a matrix, X. A subscript in parentheses represents an order statistic. A superscript in parentheses represents an

193

iteration, for example, $x_i^{(k)}$ may represent the value of x_i at the kth step of an iterative process.

x_i	The ith element of a structure (including a sample, which is a multiset).
$x_{(i)}$	The ith order statistic.
$x^{(i)}$	The value of x at the ith iteration.

Realizations of random variables and placeholders in functions associated with random variables are usually represented by lowercase letters corresponding to the uppercase letters; thus, ϵ may represent a realization of the random variable E.

A single symbol in an italic font is used to represent a single variable. A roman font or a special font is often used to represent a standard operator or a standard mathematical structure. Sometimes a string of symbols in a roman font is used to represent an operator (or a standard function), for example, exp represents the exponential function; but a string of symbols in an italic font on the same baseline should be interpreted as representing a composition (probably by multiplication) of separate objects, for example, exp represents the product of e, x, and p.

A fixed-width font is used to represent computer input or output, for example,

```
a = bx + sin(c).
```

In computer text, a string of letters or numerals with no intervening spaces or other characters, such as `bx` above, represents a single object, and there is no distinction in the font to indicate the type of object.

Some important mathematical structures and other objects are:

$\mathbb{R}$	The field of reals, or the set over which that field is defined.
$\mathbb{R}^d$	The usual d-dimensional vector space over the reals, or the set of all d-tuples with elements in $\mathbb{R}$.
$\mathbb{Z}$	The ring of integers, or the set over which that ring is defined.
$\mathbb{G}(n)$	A Galois field defined on a set with n elements.

Notation Relating to Random Variables

A common function used with continuous random variables is a *density function*, and a common function used with discrete random variables is a *probability function*. The more fundamental function for either type of random variable is the *cummulative distribution function*, or CDF. The CDF of a random variable X, denoted by $P_X(x)$, or just by $P(x)$, is defined by

$$P(x) = \text{probability}(X \le x),$$

where "probability" can be taken here as a primitive (it is defined in terms of a measure) and for vectors (of the same length), "$X \le x$" means that each element of X is less than or equal to the corresponding element of x. Both the CDF and the density or probability function for a d-dimensional random variable are defined over $\mathbb{R}^d$. (It is unfortunately necessary to state that "$P(x)$" means the "function P evaluated at x"; and likewise "$P(y)$" means the same "function P evaluated at y", unless P has been redefined. Using a different expression as the argument *does not redefine* the function, despite the sloppy convention adopted by some statisticians.)

The density for a continuous random variable is just the derivative of the CDF (if it exists). The CDF is therefore the integral. To keep the notation simple, we likewise consider the probablity function for a discrete random variable to be a type of derivative (a Radon-Nikodym derivative) of the CDF. Instead of expressing the CDF of a discrete random variable as a sum over a countable set, we often also express it as an integral. (In this case, however, the integral is over a set whose ordinary Lebesgue measure is 0.)

Functions and operators such as Cov and E that are commonly associated with Latin letters or groups of Latin letters are generally represented by that letter in a roman font.

$E(g(X))$	The expected value of the function g of the random variable X.
$V(g(X))$	The variance of the function g of the random variable X.
$\text{Cov}(X, Y)$	The covariance of the random variables X and Y.
$\text{Corr}(X, Y)$	The correlation of the random variables X and Y.
$E_p(\cdot)$	The expected value with respect to p, a probability density function or some other identifier of a probability distribution; similar meanings for $V_p(\cdot)$, $\text{Cov}_p(\cdot, \cdot)$, and $\text{Corr}_p(\cdot, \cdot)$.
$\text{Pr}(A)$	The probability of the event A.
$p_X(\cdot)$ or $P_X(\cdot)$	The probability density function (or probability function), and the cumulative probability function of the random variable X.

$p_{XY}(\cdot)$
or $P_{XY}(\cdot)$

The joint probability density function (or probability function), and the joint cumulative probability function of the random variables X and Y.

$p_{X|Y}(\cdot)$
or $P_{X|Y}(\cdot)$

The conditional probability density function (or probability function), and the conditional cumulative probability function of the random variable X given the random variable Y (these functions are random variables).

$p_{X|y}(\cdot)$
or $P_{X|y}(\cdot)$

The conditional probability density function (or probability function), and the conditional cumulative probability function of the random variable X given the realization y.

Sometimes the notation above is replaced by a similar notation in which the arguments indicate the nature of the distribution; for example $p(x, y)$ or $p(x|y)$.

$Y \sim D_X(\theta)$

The random variable Y is distributed as $D_X(\theta)$, where X is the name of a random variable associated with the distribution, and θ is a parameter of the distribution. The subscript may take similar forms to those used in the density and distribution functions, such as $X|y$, or it may be omitted. Alternatively, in place of D_X, a symbol denoting a specific distribution may be used. An example is $Z \sim N(0, 1)$, which means that Z has a normal distribution with mean 0 and variance 1.

CDF

A cumulative distribution function.

i.i.d.

Independently and identically distributed.

General Mathematical Functions and Operators

Functions such as sin, max, span, and so on that are commonly associated with groups of Latin letters are generally represented by those letters in a roman font.

Operators such as d (the differential operator) that are commonly associated with a Latin letter are generally represented by that letter in a roman font.

$\times$

Cartesian or cross product of sets, or multiplication of elements of a field or ring.

$\log x$

The natural logarithm evaluated at x.

sin x The sine evaluated at x (in radians), and similarly for other trignometric functions.

$\lceil x \rceil$ The ceiling function evaluated at the real number x: $\lceil x \rceil$ is the largest integer less than or equal to x.

$\lfloor x \rfloor$ The floor function evaluated at the real number x: $\lfloor x \rfloor$ is the smallest integer greater than or equal to x.

$x!$ The factorial of x. If x is a positive integer, $x! = x(x-1)\cdots 2 \cdot 1$. For other values of x, except negative integers, $x!$ is often defined as

$$x! = \Gamma(x + 1)$$

$\Gamma(\alpha)$ The complete gamma function. For α not equal to a nonpositive integer,

$$\Gamma(\alpha) = \int_0^\infty t^{\alpha-1} e^{-t}\, dt.$$

We have the useful relationship, $\Gamma(\alpha) = (\alpha - 1)!$. An important argument is $\frac{1}{2}$, and $\Gamma(\frac{1}{2}) = \sqrt{\pi}$.

$\Gamma_x(\alpha)$ The incomplete gamma function:

$$\Gamma_x(\alpha) = \int_0^x t^{\alpha-1} e^{-t}\, dt.$$

$B(\alpha, \beta)$ The complete beta function:

$$B(\alpha, \beta) = \int_0^1 t^{\alpha-1}(1 - t)^{\beta-1}\, dt,$$

where $\alpha > 0$ and $\beta > 0$. A useful relationship is

$$B(\alpha, \beta) = \frac{\Gamma(\alpha)\Gamma(\beta)}{\Gamma(\alpha + \beta)}.$$

$B_x(\alpha, \beta)$ The incomplete beta function:

$$B_x(\alpha, \beta) = \int_0^x t^{\alpha-1}(1 - t)^{\beta-1}\, dt.$$

$\oplus$ Bitwise binary exclusive-or (see page 27).

$O(f(n))$ Big O; $g(n) = O(f(n))$ means $g(n)/f(n) \to c$ as $n \to \infty$, where c is a nonzero finite constant.

$o(f(n))$ Little o; $g(n) = o(f(n))$ means $g(n)/f(n) \to 0$ as $n \to \infty$.

$o_P(f(n))$ Convergent in probability; $X(n) = o_P(f(n))$ means that for any positive ϵ, $\Pr(|X(n) - f(n)| > \epsilon) \to 0$ as $n \to \infty$.

d   The differential operator.

δ A perturbation operator; δx represents a perturbation of x, and not a multiplication of x by δ, even if x is a type of object for which a multiplication is defined.

$\Delta(\cdot, \cdot)$ A real-valued difference function; $\Delta(x, y)$ is a measure of the difference of x and y; for simple objects, $\Delta(x, y) = |x - y|$; for more complicated objects, a subtraction operator may not be defined, and Δ is a generalized difference.

$\tilde{x}$ A perturbation of the object x; $\Delta(x, \tilde{x}) = \delta x$.

$\tilde{x}$ An average of a sample of objects generically denoted by x.

$\bar{x}$ The mean of a sample of objects generically denoted by x.

x^- The multiplicative inverse of x with respect to some modulus (see page 24).

Appendix B

Solutions and Hints for Selected Exercises

1.2. With $a = 17$, the correlation of pairs of successive numbers should be about 0.09, and the plot should show 17 lines. With $a = 85$, the correlation of lag 1 is about 0.03, but the correlation of lag 2 is about -0.09.

1.5. 35 planes for 65 541 and 15 planes for 65 533.

1.7. 950 706 376, 129 027 171, 1 728 259 899, 365 181 143, 1 966 843 080, 1 045 174 992, 636 176 783, 1 602 900 997, 640 853 092, 429 916 489.

1.10. We seek x_0 such that

$$16\,807x_0 - (2^{31} - 1)c_1 = 2^{31} - 2$$

for some integer c_1. First, observe that $2^{31} - 2$ is equivalent to -1, so we use Euler's method (see, e.g., Shockley, 1967) with that simpler value, and write

$$
\begin{aligned}
x_0 &= \frac{((2^{31} - 1)c_1 - 1)}{16\,807} \\
&= 127\,773c_1 + \frac{(2836c_1 - 1)}{16\,807}
\end{aligned}
$$

Because the latter term must also be an integer, we write

$$16\,807c_2 = 2836c_1 - 1$$

or

$$c_1 = 5c_2 + \frac{(2627c_2 + 1)}{2836}$$

199

for some integer c_2. Continuing,

$$c_2 = c_3 + \frac{(209c_3 - 1)}{2627}$$

$$c_3 = 12c_4 + \frac{(119c_4 + 1)}{209}$$

$$c_4 = c_5 + \frac{(90c_5 - 1)}{119}$$

$$c_5 = c_6 + \frac{(29c_6 + 1)}{90}$$

$$c_6 = 3c_7 + \frac{(3c_7 - 1)}{29}$$

$$c_7 = 9c_8 + \frac{(3c_8 + 1)}{3}$$

$$c_8 = c_9 + \frac{(c_9 - 1)}{2}$$

Letting $c_9 = 1$, we can backsolve to get $x_0 = 739\,806\,647$.

1.11. Using Maple, for example,

```
> pr := 0:
> while pr < 8191 do
>     pr := primroot(pr, 8191)
> od;
```

yields the 1728 primitive roots, starting with the smallest one, 17, and going through the largest, 8180. To use `primroot`, you may first have to attach the number theory package: `with(numtheory):`.

1.12. 0.5.

2.1a. X is a random variable with an absolutely continuous distribution function P. Let Y be the random variable $P(X)$. Then, for $0 \le t \le 1$, using the existence of P^{-1},

$$
\begin{aligned}
\Pr(Y \le t) &= \Pr(P(X) \le t) \\
&= \Pr(X \le P^{-1}(t)) \\
&= P^{-1}(P(t)) \\
&= t.
\end{aligned}
$$

Hence, Y has a $U(0,1)$ distribution.

2.2. Let Z be the random variable delivered. For any x, because Y (from the density g) and U are independent, we have

$$\Pr(Z \le x) = \Pr\left(Y \le x \,\middle|\, U \le \frac{p(Y)}{cg(Y)}\right)$$

$$= \frac{\int_{-\infty}^{x} \int_{0}^{p(t)/cg(t)} g(t) \, ds \, dt}{\int_{-\infty}^{\infty} \int_{0}^{p(t)/cg(t)} g(t) \, ds \, dt}$$

$$= \int_{-\infty}^{x} p(t) \, dt,$$

the distribution function corresponding to p. Differentiating this quantity with respect to x yields $p(x)$.

2.4a. Using the relationship,

$$\frac{1}{\sqrt{2\pi}} e^{-\frac{x^2}{2}} \leq \frac{1}{\sqrt{2\pi}} e^{\frac{1}{2} - |x|}$$

(see Devroye, 1986), we have the following algorithm, after simplification:

1. Generate x from the double exponential and generate u from $U(0,1)$.
2. If $x^2 + 1 - 2|x| \leq -2 \log u$, then deliver x; otherwise go to step 1.

2.5. As $x \to \infty$, there is no c such that $cg(x) \geq p(x)$, where g is the normal density and f is the exponential density.

2.7a. $E(T) = c$; $V(T) = c^2 - c$. (Note that $c \geq 1$.)

2.8. For any t, we have

$$
\begin{aligned}
\Pr(X \leq t) &= \Pr(X \leq s + rh) \quad (\text{for } 0 \leq r \leq 1) \\
&= \Pr(U \leq r \mid V \leq U + p(s + hU)/b) \\
&= \frac{\int_{0}^{r} \int_{u}^{u+p(s+hu)/b} 2 \, dv \, du}{\int_{0}^{1} \int_{u}^{u+p(s+hu)/b} 2 \, dv \, du} \\
&= \frac{\int_{0}^{r} (p(s + hu)/b) \, du}{\int_{0}^{1} (p(s + hu)/b) \, du} \\
&= \int_{s}^{t} p(x) \, dx,
\end{aligned}
$$

where all of the symbols correspond to those in Algorithm 2.5, with the usual convention of upper case representing random variables and lower case representing constants or realizations of random variables.

3.3. Use the fact that U and $1 - U$ have the same distribution.

3.6b. A simple program using the IMSL routine `bnrdf` can be used to compute r. Here is a fragment of code that will work:

```
10 pl = bnrdf(z1,z2,rl)
   ph = bnrdf(z1,z2,rh)
```

```
        if (abs(ph-pl) .le. eps) go to 99
        rt = rl + (rh-rl)/2.
        pt = bnrdf(z1,z2,rt)
        if (pt .gt. prob) then
            rh = rt
        else
            rl = rt
        endif
        go to 10
 99     continue
        print *, rl
```

5.2b. The set is a random sample from the distribution with density f.

5.4b. The optimum is $l = r$.

5.4d. An unbiased estimator for θ is

$$\frac{d^2(n_1 + n_2)}{(dl - l^2)n}.$$

The optimum is $l = r$.

5.11a.

$$
\begin{aligned}
\mathrm{E}_{\widehat{P}}(\bar{x}_b^*) &= \mathrm{E}_{\widehat{P}}\left(\frac{1}{n}\sum_i x_i^*\right) \\
&= \frac{1}{n}\sum_i \mathrm{E}_{\widehat{P}}(x_i^*) \\
&= \frac{1}{n}\sum_i \bar{x} \\
&= \bar{x}
\end{aligned}
$$

Note that the empirical distribution is a conditional distribution, given the sample. With the sample fixed, $\bar{x}$ is a "parameter", rather than a "statistic".

5.11b.

$$
\begin{aligned}
\mathrm{E}_P(\bar{x}_b^*) &= \mathrm{E}_P\left(\frac{1}{n}\sum_i x_i^*\right) \\
&= \frac{1}{n}\sum_i \mathrm{E}_P(x_i^*) \\
&= \frac{1}{n}\sum_i \mu \\
&= \mu
\end{aligned}
$$

Alternatively,

$$
\begin{aligned}
E_P(\bar{x}_b^*) &= E_P\big(E_{\widehat{P}}(\bar{x}_b^*)\big) \\
&= E_P(\bar{x}) \\
&= \mu.
\end{aligned}
$$

5.11c. First, note

$$E_{\widehat{P}}(\overline{\bar{x}_b^*}) = \bar{x},$$

$$V_{\widehat{P}}(\bar{x}_b^*) = \frac{1}{n}\frac{1}{n}\sum(x_i - \bar{x})^2$$

and

$$V_{\widehat{P}}(\overline{\bar{x}_b^*}) = \frac{1}{Bn^2}\sum(x_i - \bar{x})^2.$$

Now,

$$
\begin{aligned}
E_{\widehat{P}}(V) &= \frac{1}{B-1}E_{\widehat{P}}\left(\sum_b\left(\bar{x}_b^* - \overline{\bar{x}_b^*}\right)^2\right) \\
&= \frac{1}{B-1}E_{\widehat{P}}\left(\sum_b \bar{x}_b^{*2} - B\overline{\bar{x}_b^*}^2\right) \\
&= \frac{1}{B-1}\left(B\bar{x}^2 + \frac{B}{n}\sum(x_i - \bar{x})^2 - B\bar{x}^2 - \frac{B}{Bn^2}\sum(x_i - \bar{x})^2\right) \\
&= \frac{1}{B-1}\left(\frac{B}{n^2}\sum(x_i - \bar{x})^2 - \frac{1}{n^2}\sum(x_i - \bar{x})^2\right) \\
&= \frac{1}{n^2}\sum(x_i - \bar{x})^2 \\
&= \frac{1}{n}\sigma_{\widehat{P}}^2
\end{aligned}
$$

5.11d.

$$
\begin{aligned}
E_P(V) &= E_P(E_{\widehat{P}}(V)) \\
&= E_P\left(\frac{1}{n}\sum(x_i - \bar{x})^2 n\right) \\
&= \frac{1}{n}\frac{n-1}{n}\sigma_P^2
\end{aligned}
$$

5.7a. Generate x_i from a gamma(3,2) distribution, and take your estimator as

$$16\frac{\sum \sin(\pi x_i)}{n}.$$

6.2. Out of the 100 trials, 97 times the maximum element is in position 1311. The test is really not valid because the seeds are all relatively small and are very close together. Try the same test, but with 100 randomly generated seeds.

Bibliography

As might be expected, the literature in the interface of computer science, numerical analysis, and statistics is quite diverse; and articles on random number generation and Monte Carlo methods are likely to appear in journals devoted to quite different disciplines. There are at least ten journals and serials whose titles contain some variants of both "computing" and "statistics"; but there are far more journals in numerical analysis and in areas such as "computational physics", "computational biology", and so on that publish articles relevant to the fields of statistical computing and computational statistics. Many of the methods of computational statistics involve random number generation and Monte Carlo methods. The journals in the mainstream of statistics also have a large proportion of articles in the fields of statistical computing and computational statistics because, as we suggested in the preface, recent developments in statistics and in the computational sciences have paralleled each other to a large extent.

There are two well-known learned societies whose primary focus is in statistical computing: the International Association for Statistical Computing (IASC), which is an affiliated society of the International Statistical Institute, and the Statistical Computing Section of the American Statistical Association (ASA). The Statistical Computing Section of the ASA has a regular newsletter carrying news and notices as well as articles on practicum. The activities of the Society for Industrial and Applied Mathematics (SIAM) are often relevant to computational statistics.

There are two regular conferences in the area of computational statistics: COMPSTAT, held biennially in Europe and sponsored by the IASC, and the Interface Symposium, generally held annually in North America and sponsored by the Interface Foundation of North America with cooperation from the Statistical Computing Section of the ASA.

In addition to literature and learned societies in the traditional forms, an important source of communication and a repository of information are computer databases and forums. In some cases the databases duplicate what is available in some other form, but often the material and the communications facilities provided by the computer are not available elsewhere.

Literature in Computational Statistics

In the Library of Congress classification scheme, most books on statistics, including statistical computing, are in the QA276 section, although some are classified under H, HA, and HG. Numerical analysis is generally in QA279, and computer science in QA76. Many of the books in the interface of these disciplines are classified in these or other places within QA.

Current Index to Statistics, published annually by the American Statistical Association and the Institute for Mathematical Statistics, contains both author and subject indexes that are useful in finding journal articles or books in statistics. The *Index* is available in hard copy and on CD-ROM. The CD-ROM version with software developed by Ron Thisted and Doug Bates is particularly useful. In passing, I take this opportunity to acknowledge the help this database and software were to me in tracking down references for this book.

The Association for Computing Machinery (ACM) publishes an annual index, by author, title, and keyword, of the literature in the computing sciences.

Mathematical Reviews, published by the American Mathematical Society (AMS), contains brief reviews of articles in all areas of mathematics. The areas of "Statistics", "Numerical Analysis", and "Computer Science" contain reviews of articles relevant to computational statistics. The papers reviewed in *Mathematical Reviews* are categorized according to a standard system that has slowly evolved over the years. In this taxonomy, called the AMS MR classification system, "Statistics" is 62Xyy; "Numerical Analysis", including random number generation, is 65Xyy; and "Computer Science" is 68Xyy. ("X" represents a letter and "yy" represents a two-digit number.)

Mathematical Reviews is also available to subscribers via the World Wide Web at MathSciNet.

There are various handbooks of mathematical functions and formulas that are useful in numerical computations. Three that should be mentioned are Abramowitz and Stegun (1964), Spanier and Oldham (1987) and Thompson (1997). Anyone doing serious scientific computations should have ready access to at least one of these volumes.

Almost all journals in statistics have occasional articles on computational statistics and statistical computing. The following is a list of journals and proceedings that emphasize this field.

ACM Transactions on Mathematical Software, published quarterly by the ACM (Association for Computing Machinery). (Includes algorithms in Fortran and C. Most of the algorithms are available through `netlib`. The ACM collection of algorithms is sometimes called *CALGO*.)

ACM Transactions on Modeling and Computer Simulation, published quarterly by the ACM.

Applied Statistics, published quarterly by the Royal Statistical Society. (Includes algorithms in Fortran. Most of the algorithms are available through `statlib` at Carnegie Mellon University. Some of these algorithms, with corrections, were collected by Griffiths and Hill, 1985)

Communications in Statistics — Simulation and Computation, published quarterly by Marcel Dekker. (Includes algorithms in Fortran. Until 1982, this journal was designated as *Series B*.)

Computational Statistics, published quarterly by Physica-Verlag. (Formerly called *Computational Statistics Quarterly*.)

Computational Statistics. Proceedings of the xxth Symposium on Computational Statistics (COMPSTAT), published biennially by Physica-Verlag. (Not refereed.)

Computational Statistics and Data Analysis, published quarterly by North Holland. (This is also the official journal of the International Association for Statistical Computing.)

Computing Science and Statistics. This is an annual publication containing papers presented at the Interface Symposium. Until 1992, these proceedings were named *Computer Science and Statistics: Proceedings of the xxth Symposium on the Interface*. (The 24th symposium was held in 1992.) These proceedings are now published by the Interface Foundation of North America. (Not refereed.)

Journal of Computational and Graphical Statistics, published quarterly by the American Statistical Association.

Journal of Statistical Computation and Simulation, published quarterly by Gordon Breach.

Proceedings of the Statistical Computing Section, published annually by the American Statistical Association. (Not refereed.)

SIAM Journal on Scientific Computing, published bimonthly by SIAM. This journal was formerly *SIAM Journal on Scientific and Statistical Computing*. (Is this a step backward?)

Statistics and Computing, published quarterly by Chapman & Hall.

There are two journals whose contents are primarily in the subject area of random number generation, simulation, and Monte Carlo methods: *ACM Transactions on Modeling and Computer Simulation* (Volume 1 appeared in 1992) and *Monte Carlo Methods and Applications* (Volume 1 appeared in 1995).

There has been a series of conferences concentrating on this area (with an emphasis on quasi-random methods). The first International Conference on Monte Carlo and Quasi-Monte Carlo Methods in Scientific Computing was held in Las Vegas, Nevada, in 1994; the second in Salzburg, Austria, in 1996; and the third in Claremont, California, in 1998. The proceedings of the conferences have been published in the Lecture Notes in Statistics series of Springer-Verlag. The proceedings of the first conference were published as Niederreiter and Shiue (1995), and those of the second as Niederreiter, Hellekalek, Larcher, and Zinterhof (1997).

The proceedings of the CRYPTO conferences often contain interesting articles on uniform random number generation, with an emphasis on the cryptographic applications. These proceedings are published in the Lecture Notes in Computer Science series of Springer-Verlag under the name *Proceedings of*

CRYPTO XX, where XX is a two digit number representing the year of the conference.

There are a number of textbooks, monographs, and survey articles on random number generation and Monte Carlo methods. Some of particular note are Bratley, Fox, and Schrage (1987); Dagpunar (1988); Deák (1990); Devroye (1986); Fishman (1996); Knuth (1981); L'Ecuyer (1990); Lewis and Orav (1989); Morgan (1984); Niederreiter (1992, 1995c); Ripley (1987); and Tezuka (1995).

World Wide Web, News Groups, List Servers, and Bulletin Boards

The best way of storing information is in a digital format that can be accessed by computers. In some cases the best way for people to access information is by computers; in other cases the best way is via hard copy, which means that the information stored on the computer must go through a printing process resulting in books, journals, or loose pages.

A huge amount of information and raw data is available online. Much of it is in publicly accessible sites. Some of the repositories give space to ongoing discussions to which anyone can contribute.

There are various ways of remotely accessing the computer databases and discussion groups. The high-bandwidth wide-area network called the "Internet" is the most important way to access information. Early development of the Internet was due to initiatives within the United States Department of Defense and the National Science Foundation. The Internet is making fundamental changes to the way we store and access information.

The references that I have cited in this text are generally traditional books, journal articles, or compact disks. This usually means that the material has been reviewed by someone other than the author. It also means that the author possibly has newer thoughts on the same material. The Internet provides a mechanism for the dissemination of large volumes of information that can be updated readily. The ease of providing material electronically is also the source of the major problem with the material: it is often half-baked and has not been reviewed critically. Another reason that I have refrained from making frequent reference to material available over the Internet is the unreliability of some sites. It has been estimated that the average life of a Web site is 45 days (in early 1998).

The World Wide Web (WWW)

Mechanisms to access information from various sites on the Internet have been developed, beginning with early work at CERN in Switzerland. The development of the Mosaic Web browser at the National Center for Supercomputer

Applications at the University of Illinois marked a major turning point in ease-of-access of information over the Internet.

The Web browsers rely on standard ways of formatting text and images and of linking sites. These methods are independent of the Internet; indeed, they are useful on a single computer for developing an information access system. The basic coding schemes are incorporated in `html` and `xml`. The main new facility provided by `xml` are content tags, which allow specification of the meanings of markings, thus facilitating searches. The World Wide Web Consortium (W3C) povides directions and promotes standards for these markup languages. See

> `http://www.w3.org/`

A very important extension of `xml` is `mathml`, which provides special markups and tags for mathematical objects.

Actions can be initiated remotely over the Web, using programming languages such as Java.

For linking Internet sites, the Web browsers use a "Universal Resource Locator" (URL), which determines the location and the access method. "URL" is also used to mean the method and the site ("location", instead of "locator"); for example,

> `http://www.science.gmu.edu/~jgentle/rngbk`

is called a URL. (As mentioned in the preface, this is the URL for a site that I maintain to store information about this book.)

For statistics, one of the most useful sites on the Internet is the electronic repository `statlib`, maintained at Carnegie Mellon University, which contains programs, datasets, and other items of interest. The URL is

> `http://lib.stat.cmu.edu`.

The collection of algorithms published in *Applied Statistics* is available in `statlib`. These algorithms are sometimes called the *ApStat* algorithms.

The `statlib` facility can also be accessed by email or anonymous `ftp` at

> `statlib@temper.stat.cmu.edu`.

An automatic email processor will reply with files that provide general information or programs and data. The general introductory file can be obtained by sending email to the address above with the message "`send index`".

Another very useful site for scientific computing is `netlib`, which was established by research workers at AT&T (now Lucent) Bell Laboratories and national laboratories, primarily Oak Ridge National Laboratories. The URL is

> `http://www.netlib.org`

The collection of ACM algorithms (*CALGO*) is available in `netlib`.

There is also an X Windows, socket-based system for accessing `netlib`, called `Xnetlib`; see Dongarra, Rowan, and Wade (1995).

The Guide to Available Mathematical Software (GAMS), to which I have referred several times in this book, can be accessed at

```
http://gams.nist.gov
```

A different interface, using Java, is available at

```
http://math.nist.gov/HotGAMS/
```

There are two major problems in using the WWW to gather information. One is the sheer quantity of information and the number of sites providing information. The other is the "kiosk problem"; anyone can put up material. Sadly, the average quality is affected by a very large denominator. The kiosk problem may be even worse than a random selection of material; the "fools in public places" syndrome is much in evidence.

There is not much that can be done about the second problem. It was not solved for traditional postings on uncontrolled kiosks, and it will not be solved on the WWW.

For the first problem, there are remarkable programs that automatically crawl through WWW links to build a database that can be searched for logical combinations of terms and phrases. Such systems and databases have been built by several people and companies.

Two of the most useful are Alta Vista, provided by Digital Equipment Corporation, at

```
http://www.altavista.digital.com
```

and HotBot at

```
http://www.hotbot.com
```

In a study by Lawrence and Giles (1998), these two full-text search engines provided far more complete coverage of the scientific literature than four other search engines considered. The Lawrence and Giles study indicated that use of all six search engines provided about 3.5 times as many documents on average as use of just a single engine.

A very widely used search program is "Yahoo" at

```
http://www.yahoo.com
```

A neophyte can be quickly disabused of an exaggerated sense of the value of such search engines by doing a search on "Monte Carlo". Aside from the large number of hits that relate to a car and to some place in Europe, the hits (in mid 1998) that relate to the interesting topic are dominated by references to some programs for random number generation put together by a group at a university somewhere. (Of course, "interesting" is in the eye of the beholder.)

It is not clear at this time what will be the media for the scientific literature within a few years. Many of the traditional journals will be converted to an electronic version of some kind. Journals will become Web sites. That is for certain; the details, however, are much less certain. Many bulletin boards and discussion groups have already evolved into "electronic journals'. A publisher of a standard commercial journal has stated that "we reject 80% of the articles submitted to our journal; those are the ones you can find on the Web".

Lesk (1997) discusses many of the issues that must be considered as the standard repositories of knowledge change from paper books and journals to digital libraries.

References

The following bibliography obviously covers a wide range of topics in random number generation and Monte Carlo methods. Except for a few of the general references, all of these entries have been cited in the text.

The purpose of this bibliography is to help the reader get more information; hence I eschew "personal communications" and references to technical reports that may or may not exist. Those kinds of references are generally for the author rather than for the reader.

A Note on the Names of Authors

In these references, I have generally used the names of authors as they appear in the original sources. This may mean that the same author will appear with different forms of names, sometimes with given names spelled out, and sometimes abbreviated. In the author index, beginning on page 237, I use a single name for the same author. The name is generally the most unique (i.e., least abbreviated) of any of the names of that author in any of the references. This convention may occasionally result in an entry in the author index that does not occur exactly in any references. A reference to J. Paul Jones together with one to John P. Jones, if I know that the two names refer to the same person, would result in an Author Index entry for John Paul Jones.

Abramowitz, Milton, and Irene A. Stegun (Editors) (1964), *Handbook of Mathematical Functions with Formulas, Graphs, and Mathematical Tables*, National Bureau of Standards (NIST), Washington. (Reprinted by Dover Publications, Inc., New York.)

Afflerbach, L., and H. Grothe (1985), Calculation of Minkowski-reduced lattice bases, *Computing* **35**, 269–276.

Afflerbach, Lothar, and Holger Grothe (1988), The lattice structure of pseudorandom vectors generated by matrix generators, *Journal of Computational and Applied Mathematics* **23**, 127–131.

Afflerbach, L., and W. Hörmann (1992), Nonuniform random numbers: A sensitivity analysis for transformation methods, *International Workshop on Computationally Intensive Methods in Simulation and Optimization* (edited by U. Dieter and G. C. Pflug), Springer-Verlag, Berlin, 374.

Agresti, Alan (1992), A survey of exact inference for contingency tables (with discussion), *Statistical Science* **7**, 131–177.

Ahn, Hongshik, and James J. Chen (1995), Generation of over-dispersed and under-dispersed binomial variates, *Journal of Computational and Graphical Statistics* **4**, 55–64.

Ahrens, J. H., and U. Dieter (1972), Computer methods for sampling from the exponential and normal distributions, *Communications of the ACM* **15**, 873–882.

Ahrens, J. H., and U. Dieter (1974), Computer methods for sampling from gamma, beta, Poisson, and binomial distributions, *Computing* **12**, 223–246.

Ahrens, J. H., and U. Dieter (1980), Sampling from binomial and Poisson distributions: A method with bounded computation times, *Computing* **25**, 193–208.

Ahrens, J. H., and U. Dieter (1985), Sequential random sampling, *ACM Transactions on Mathematical Software* **11**, 157–169.

Ahrens, Joachim H., and Ulrich Dieter (1988), Efficient, table-free sampling methods for the exponential, Cauchy and normal distributions, *Communications of the ACM* **31**, 1330–1337.

Ahrens, J. H., and U. Dieter (1991), A convenient sampling method with bounded computation times for Poisson distributions, *The Frontiers of Statistical Computation, Simulation & Modeling* (edited by P. R. Nelson, E. J. Dudewicz, A. Öztürk, and E. C. van der Meulen), American Sciences Press, Columbus, Ohio, 137–149.

Akima, Hirosha (1970), A new method of interpolation and smooth curve fitting based on local procedures, *Journal of the ACM* **17**, 589–602.

Alonso, Laurent, and René Schott (1995), *Random Generation of Trees: Random Generators in Science*, Kluwer Academic Publishers, Boston.

Altman, N. S. (1989), Bit-wise behavior of random number generators, *SIAM Journal on Scientific and Statistical Computing* **9**, 941–949.

Aluru, S.; G. M. Prabhu; and John Gustafson (1992), A random number generator for parallel computers, *Parallel Computing* **18**, 839–847.

Anderson, N. H., and D. M. Titterington (1993), Cross-correlation between simultaneously generated sequences of pseudo-random uniform deviates, *Statistics and Computing* **3**, 61–65.

Anderson, S. L. (1990), Random number generators on vector supercomputers and other advanced architectures, *SIAM Review* **32**, 221–251.

Anderson, T. W.; I. Olkin; and L. G. Underhill (1987), Generation of random orthogonal matrices, *SIAM Journal on Scientific and Statistical Computing* **8**, 625–629.

Andrews, D. F.; P. J. Bickel; F. R. Hampel; P. J. Huber; W. H. Rogers; and J. J. Tukey (1972), *Robust Estimation of Location: Survey and Advances*, Princeton University Press, Princeton.

Antonov, I. A., and V. M. Saleev (1979), An economic method of computing LP_τ-sequences, *USSR Computational Mathematics and Mathematical Physics* **19**, 252–256.

Arnason, A. N., and L. Baniuk (1978), A computer generation of Dirichlet variates, *Proceedings of the Eighth Manitoba Conference on Numerical Mathematics and Computing*, Utilitas Mathematica Publishing, Winnipeg, 97–105.

Arnold, Barry C. (1983), *Pareto Distributions*, International Co-operative Publishing House, Fairland, Maryland.

Atkinson, A. C. (1979), A family of switching algorithms for the computer generation of beta random variates, *Biometrika* **66**, 141–145.

Atkinson, A. C. (1980), Tests of pseudo-random numbers. *Applied Statistics* **29**, 164–171.

Avramidis, Athanassios N., and James R. Wilson (1995), Correlation-induction techniques for estimating quantiles in simulation experiments, *Proceedings of the 1995 Winter Simulation Conference*, Association for Computing Machinery, New York, 268–277.

Bailey, Ralph W. (1994), Polar generation of random variates with the *t*-distribution, *Mathematics of Computation* **62**, 779–781.

Balakrishnan, N., and R. A. Sandhu (1995), A simple simulation algorithm for generating progressive Type-II censored samples, *The American Statistician* **49**, 229–230.

Banerjia, Sanjeev, and Rex A. Dwyer (1993), Generating random points in a ball, *Communications in Statistics — Simulation and Computation* **22**, 1205–1209.

Barnard, G. A. (1963), Discussion of Bartlett, "The spectral analysis of point processes", *Journal of the Royal Statistical Society, Series B* **25**, 264–296.

Bays, Carter, and S. D. Durham (1976), Improving a poor random number generator, *ACM Transactions on Mathematical Software* **2**, 59–64.

Beck, J., and W. W. L. Chen (1987), *Irregularities of Distribution*, Cambridge University Press, Cambridge, United Kingdom.

Becker, P. J., and J. J. J. Roux (1981), A bivariate extension of the gamma distribution, *South African Statistical Journal* **15**, 1–12.

Beckman, Richard J., and Michael D. McKay (1987), Monte Carlo estimation under different distributions using the same simulation, *Technometrics* **29**, 153–160.

Bélisle, Claude J. P.; H. Edwin Romeijn; and Robert L. Smith (1993), Hit-and-run algorithms for generating multivariate distributions, *Mathematics of Operations Research* **18**, 255–266.

Bendel, R. B., and M. R. Mickey (1978), Population correlation matrices for sampling experiments, *Communications in Statistics — Simulation and Computation* **B7**, 163–182.

Berbee, H. C. P.; C. G. E. Boender; A. H. G. Rinnooy Kan; C. L. Scheffer; R. L. Smith; and J. Telgen (1987), Hit-and-run algorithms for the identification of nonredundant linear inequalities, *Mathematical Programming* **37**, 184–207.

Best, D. J. (1983), A note on gamma variate generators with shape parameter less than unity, *Computing* **30**, 185–188.

Best, D. J., and N. I. Fisher (1979), Efficient simulation of the von Mises distribution, *Applied Statistics* **28**, 152–157.

Beyer, W. A. (1972), Lattice structure and reduced bases of random vectors generated by linear recurrences, *Applications of Number Theory to Numerical Analysis* (edited by S. K. Zaremba), Academic Press, New York, 361–370.

Beyer, W. A.; R. B. Roof; and D. Williamson (1971), The lattice structure of multiplicative congruential pseudo-random vectors, *Mathematics of Computation* **25**, 345–363.

Bhanot, Gyan (1988), The Metropolis algorithm, *Reports on Progress in Physics* **51**, 429–457.

Birkes, David, and Yadolah Dodge (1993), *Alternative Methods of Regression*, John Wiley & Sons, New York.

Blum, L.; M. Blum; and M. Shub (1986), A simple unpredictable pseudo-random number generator, *SIAM Journal of Computing* **15**, 364–383.

Bouleau, Nicolas, and Dominique Lépingle (1994), *Numerical Methods for Stochastic Processes*, John Wiley & Sons, New York.

Bowman, K. O., and M. T. Robinson (1987), Studies of random number generators for parallel processing, *Proceedings of the Second Conference on Hypercube Multiprocessors*, Society for Industrial and Applied Mathematics, Philadelphia, 445–453.

Boyar, J. (1989), Inferring sequences produced by pseudo-random number generators, *Journal of the ACM* **36**, 129–141.

Boyett, J. M. (1979), Random $R \times C$ tables with given row and column totals, *Applied Statistics* **28**, 329–332.

Braaten, E., and G. Weller (1979), An improved low-discrepancy sequence for multidimensional quasi-Monte Carlo integration, *Journal of Computational Physics* **33**, 249–258.

Bratley, Paul, and Bennett L. Fox (1988), Algorithm 659: Implementing Sobol's quasirandom sequence generator, *ACM Transactions on Mathematical Software* **14**, 88–100.

Bratley, Paul; Bennett L. Fox; and Harald Niederreiter (1992), Implementation and tests of low-discrepancy sequences, *ACM Transactions on Modeling and Computer Simulation* **2**, 195–213.

Bratley, Paul; Bennett L. Fox; and Harald Niederreiter (1994), Algorithm 738: Programs to generate Niederreiter's low-discrepancy sequences, *ACM Transactions on Mathematical Software* **20**, 494–495.

Bratley, Paul; Bennett L. Fox; and Linus E. Schrage (1987), *A Guide to Simulation*, second edition, Springer-Verlag, New York.

Brophy, John F.; James E. Gentle; Jing Li; and Philip W. Smith (1989), Software for advanced architecture computers, *Computer Science and Statistics: Proceedings of the Twenty-first Symposium on the Interface* (edited by Kenneth Berk and Linda Malone), American Statistical Association, 116–120.

Brown, Morton B., and Judith Bromberg (1984), An efficient two-stage procedure for generating random variates from the multinomial distribution, *The American Statistician* **38**, 216–219.

Buckheit, Jonathan B., and David L. Dohoho (1995), WaveLab and reproducible research, *Wavelets and Statistics* (edited by Anestis Antoniadis and Georges Oppenheim) Springer-Verlag, New York, 55–81.

Buckle, D. J. (1995), Bayesian inference for stable distributions, *Journal of the American Statistical Association* **90**, 605–613.

Burr, I. W. (1942), Cumulative frequency functions, *Annals of Mathematical Statistics* **13**, 215–232.

Burr, Irwing W., and Peter J. Cislak (1968), On a general system of distributions. I. Its curve-shape characteristics. II. The sample median, *Journal of the American Statistical Association* **63**, 627–635.

Cabrera, Javier, and Dianne Cook (1992), Projection pursuit indices based on fractal dimension, *Computing Science and Statistics* **24**, 474–477.

Caflisch, Russel E., and Bradley Moskowitz (1995), Modified Monte Carlo methods using quasi-random sequences, *Monte Carlo and Quasi-Monte Carlo Methods in Scientific Computing* (edited by Harald Niederreiter and Peter Jau-Shyong Shiue), Springer-Verlag, New York, 1–16.

Carlin, Bradley P., and Thomas A. Louis (1996), *Bayes and Empirical Bayes Methods for Data Analysis*, Chapman & Hall, New York.

Carta, David G. (1990), Two fast implementations of the "minimal standard" random number generator, *Communications of the ACM* **33**, Number 1 (January), 87–88.

Casella, George, and Edward I. George (1992), Explaining the Gibbs sampler, *The American Statistician* **46**, 167–174.

Chalmers, C. P. (1975), Generation of correlation matrices with given eigenstructure, *Journal of Statistical Computation and Simulation* **4**, 133–139.

Chambers, J. M.; C. L. Mallows; and B. W. Stuck (1976), A method for simulating stable random variables. *Journal of the American Statistical Association* **71**, 340–344 (Corrections, 1987, *ibid.* **82**, 704, and 1988, *ibid.* **83**, 581).

Chen, H. C., and Y. Asau (1974), On generating random variates from an empirical distribution, *AIIE Transactions* **6**, 163–166.

Chen, Huifen, and Bruce W. Schmeiser (1992), Simulation of Poisson processes with trigonometric rates, *Proceedings of the 1992 Winter Simulation Conference*, Association for Computing Machinery, New York, 609–617.

Chen, Ming-Hui, and Bruce Schmeiser (1993), Performance of the Gibbs, hit-and-run, and Metropolis samplers, *Journal of Computational and Graphical Statistics* **3**, 251–272.

Chen, Ming-Hui, and Bruce W. Schmeiser (1996), General hit-and-run Monte Carlo sampling for evaluating multidimensional integrals, *Operations Research Letters* **19**, 161–169.

Cheng, R. C. H. (1978), Generating beta variates with nonintegral shape parameters, *Communications of the ACM* **21**, 317–322.

Cheng, R. C. H. (1985), Generation of multivariate normal samples with given mean and covariance matrix, *Journal of Statistical Computation and Simulation* **21**, 39–49.

Cheng, R. C. H., and G. M. Feast (1979), Some simple gamma variate generators, *Applied Statistics* **28**, 290–295.

Cheng, R. C. H., and G. M. Feast (1980), Gamma variate generators with increased shape parameter range, *Communications of the ACM* **23**, 389–393.

Chib, Siddhartha, and Edward Greenberg (1995), Understanding the Metropolis-Hasting algorithm, *The American Statistician* **49**, 327–335.

Chou, Wun-Seng, and Harald Niederreiter (1995), On the lattice test for inversive congruential pseudorandom numbers, *Monte Carlo and Quasi-Monte Carlo Methods in Scientific Computing* (edited by Harald Niederreiter and Peter Jau-Shyong Shiue), Springer-Verlag, New York, 186–197.

Coldwell, R. L. (1974), Correlational defects in the standard IBM 360 random number generator and the classical ideal gas correlational function, *Journal of Computational Physics* **14**, 223–226.

Collings, Bruce Jay (1987), Compound random number generators, *Journal of the American Statistical Association* **82**, 525–527.

Couture, R., and Pierre L'Ecuyer (1994), On the lattice structure of certain linear congruential sequences related to AWC/SWB generators, *Mathematics of Computation* **62**, 799–808.

Couture, Raymond, and Pierre L'Ecuyer (1995), Linear recurrences with carry as uniform random number generators, *Proceedings of the 1995 Winter Simulation Conference*, Association for Computing Machinery, New York, 263–267.

Coveyou, R. R., and R. D. MacPherson (1967), Fourier analysis of uniform random number generators, *Journal of the ACM* **14**, 100–119.

Cuccaro, Steven A.; Michael Mascagni; and Daniel V. Pryor (1994), Techniques for testing the quality of parallel pseudorandom number generators, *Proceedings of the Seventh SIAM Conference on Parallel Processing for Scientific Computing*, Society for Industrial and Applied Mathematics, Philadelphia, 279–284.

Currin, Carla; Toby J. Mitchell; Max Morris; and Don Ylvisaker (1991), Bayesian prediction of deterministic functions, with applications to the design and analysis of computer experiments, *Journal of the American Statistical Association* **86**, 953–963.

Dagpunar, John (1988), *Principles of Random Variate Generation*, Clarendon Press, Oxford, United Kingdom.

Dagpunar, J. (1990), Sampling from the von Mises distribution via a comparison of random numbers, *Journal of Applied Statistics* **17**, 165–168.

Damien, Paul; Purushottam W. Laud; and Adrian F. M. Smith (1995), Approximate random variate generation from infinitely divisible distributions with applications to Bayesian inference, *Journal of the Royal Statistical Society, Series B* **57**, 547–563.

David, Herbert A. (1981), *Order Statistics*, second edition, John Wiley & Sons, New York.

Davis, Charles S. (1993), The computer generation of multinomial random variates, *Computational Statistics and Data Analysis* **16**, 205–217.

Davison, A. C., and D. V. Hinkley (1997), *Bootstrap Methods and Their Application*, Cambridge University Press, Cambridge, United Kingdom.

Deák, I. (1981), An economical method for random number generation and a normal generator, *Computing* **27**, 113–121.

Deák, I. (1986), The economical method for generating random samples from discrete distributions, *ACM Transactions on Mathematical Software* **12**, 34–36.

Deák, István (1990), *Random Number Generators and Simulation*, Akadémiai Kiadó, Budapest.

Dellaportas, P., and A. F. M. Smith (1993), Bayesian inference for generalized linear and proportional hazards models via Gibbs sampling, *Applied Statistics* **42**, 443–459.

De Matteis, A., and S. Pagnutti (1990), Long-range correlations in linear and non-linear random number generators, *Parallel Computing* **14**, 207–210.

De Matteis, A., and S. Pagnutti (1993), Long-range correlation analysis of the Wichmann-Hill random number generator, *Statistics and Computing* **3**, 67–70.

Devroye, Luc (1986), *Non-Uniform Random Variate Generation*, Springer-Verlag, New York.

Devroye, Luc (1987), A simple generator for discrete log-concave distributions, *Computing* **39**, 87–91.

Devroye, Luc; Peter Epstein; and Jörg-Rüdiger Sack (1993), On generating random intervals and hyperrectangles, *Journal of Computational and Graphical Statistics* **2**, 291–308.

Dieter, U. (1975), How to calculate shortest vectors in a lattice, *Mathematics of Computation* **29**, 827–833.

Do, Kim-Anh (1991), Quasi-random resampling for the bootstrap, *Computer Science and Statistics: Proceedings of the Twenty-third Symposium on the Interface* (edited by Elaine M. Keramidas), Interface Foundation of North America, 297–300.

Dongarra, Jack; Tom Rowan; and Reed Wade (1995), Software distribution using `Xnetlib`, *ACM Transactions on Mathematical Software* **21**, 79–88.

Eddy, William F. (1990), Random number generators for parallel processors, *Journal of Computational and Applied Mathematics* **31**, 63–71.

Eddy, William F., and James E. Gentle (1985), Statistical computing: what's past is prologue, *A Celebration of Statistics* (edited by Anthony C. Atkinson and Stephen E. Fienberg), Springer-Verlag, Berlin, 233–249.

Efron, Bradley, and Robert J. Tibshirani (1993), *An Introduction to the Bootstrap*, Chapman & Hall, New York.

Eichenauer, J.; H. Grothe; and J. Lehn (1988), Marsaglia's lattice test and nonlinear congruential pseudo random number generators, *Metrika* **35**, 241–250.

Eichenauer, J., and H. Niederreiter (1988), On Marsaglia's lattice test for pseudorandom numbers, *Manuscripta Mathematica* **62**, 245–248.

Eichenauer, Jürgen, and Jürgen Lehn (1986), A non-linear congruential pseudo random number generator, *Statistische Hefte* **27**, 315–326.

Eichenauer-Herrmann, Jürgen (1995), Pseudorandom number generation by nonlinear methods, *International Statistical Review* **63**, 247–255.

Eichenauer-Herrmann, Jürgen (1996), Modified explicit inversive congruential pseudorandom numbers with power of 2 modulus, *Statistics and Computing*

6, 31–36.

Eichenauer-Herrmann, J., and H. Grothe (1989), A remark on long-range correlations in multiplicative congruential pseudorandom number generators, *Numerische Mathematik* **56**, 609–611.

Eichenauer-Herrmann, J., and H. Grothe (1990), Upper bounds for the Beyer ratios of linear congruential generators, *Journal of Computational and Applied Mathematics* **31**, 73–80.

Eichenauer-Herrmann, J., and K. Ickstadt (1994), Explicit inversive congruential pseudorandom numbers with power of 2 modulus, *Mathematics of Computation* **62**, 787–797.

Emrich, Lawrence J., and Marion R. Piedmonte (1991), A method for generating high-dimensional multivariate binary variates, *The American Statistician* **45**, 302–304.

Fang, Kai-Tai, and Run-Ze Li (1997), Some methods for generating both an NT-net and the uniform distribution on a Stiefel manifold and their applications, *Computational Statistics and Data Analysis* **24**, 29–46.

Fang, Kai-Tai, and Yuan Wang (1994), *Number Theoretic Methods in Statistics*, Chapman & Hall, New York.

Faure, H. (1986), On the star discrepancy of generalised Hammersley sequences in two dimensions. *Monatshefte für Mathematik* **101**, 291–300.

Ferrenberg, A. M.; D. P. Landau; and Y. J. Wong (1992), Monte Carlo simulations: Hidden errors from "good" random number generators, *Physical Review Letters* **69**, 3382–3384.

Fishman, George S., and Louis R. Moore, III (1982), A statistical evaluation of multiplicative random number generators with modulus $2^{31} - 1$, *Journal of the American Statistical Association* **77**, 129–136.

Fishman, George S., and Louis R. Moore, III (1986), An exhaustive analysis of multiplicative congruential random number generators with modulus $2^{31} - 1$, *SIAM Journal on Scientific and Statistical Computing* **7**, 24–45.

Fleishman, Allen I. (1978), A method for simulating non-normal distributions, *Psychometrika* **43**, 521–532.

Forster, Jonathan J.; John W. McDonald; and Peter W. F. Smith (1996), Monte Carlo exact conditional tests for log-linear and logistic models, *Journal of the Royal Statistical Society, Series B* **55**, 3–24.

Fox, Bennett L. (1986), Implementation and relative efficiency of quasirandom sequence generators, *ACM Transactions on Mathematical Software* **12**, 362–376.

Frederickson, P.; R. Hiromoto; T. L. Jordan; B. Smith; and T. Warnock (1984), Pseudo-random trees in Monte Carlo, *Parallel Computing* **1**, 175-180.

Freund, John E. (1961), A bivariate extension of the exponential distribution, *Journal of the American Statistical Association* **56**, 971–977.

Friedman, Jerome H.; Jon Louis Bentley; and Raphael Ari Finkel (1977), An algorithm for finding best matches in logarithmic expected time, *ACM Transactions on Mathematical Software* **3**, 209–226.

Frigessi, A.; F. Martinelli; and J. Stander (1997), Computational complexity of Markov chain Monte Carlo methods for finite Markov random fields, *Biometrika* **84**, 1–18.

Fuller, A. T. (1976), The period of pseudo-random numbers generated by Lehmer's congruential method, *Computer Journal* **19**, 173–177.

Fushimi, Masanori (1990), Random number generation with the recursion $X_t = X_{t-3p} \oplus X_{t-3q}$, *Journal of Computational and Applied Mathematics* **31**, 105–118.

Gamerman, Dani (1997), *Markov Chain Monte Carlo*, Chapman & Hall, London.

Gange, Stephen J. (1995), Generating multivariate categorical variates using the iterative proportional fitting algorithm, *The American Statistician* **49**, 134–138.

Gelfand, Alan E., and Adrian F. M. Smith (1990), Sampling-based approaches to calculating marginal densities, *Journal of the American Statistical Association* **85**, 398–409.

Gelfand, Alan E., and Sujit K. Sahu (1994), On Markov chain Monte Carlo acceleration, *Journal of Computational and Graphical Statistics* **3**, 261–276.

Gelman, Andrew (1992), Iterative and non-iterative simulation algorithms, *Computing Science and Statistics* **24**, 4433-438.

Gelman, Andrew, and Donald B. Rubin (1992a), Inference from iterative simulation using multiple sequences (with discussion), *Statistical Science* **7**, 457–511.

Gelman, Andrew, and Donald B. Rubin (1992b), A single series from the Gibbs sampler provides a false sense of security, *Bayesian Statistics 4* (edited by J. M. Bernardo, J. O. Berger, A. P. Dawid, and A. F. M. Smith), Oxford University Press, Oxford, United Kingdom, 625–631.

Gelman, Andrew; John B. Carlin; Hal S. Stern; and Donald B. Rubin (1995), *Bayesian Data Analysis*, Chapman & Hall, London.

Geman, S., and D. Geman (1984), Stochastic relaxation, Gibbs distributions, and the Bayesian restoration of images, *IEEE Transactions on Pattern Analysis and Machine Intelligence* **6**, 721–741.

Gentle, James E. (1981), Portability considerations for random number generators, *Computer Science and Statistics: Proceedings of the 13th Symposium on the Interface* (edited by William F. Eddy), Springer-Verlag, New York, 158–164.

Gentle, James E. (1990), Computer implementation of random number generators, *Journal of Computational and Applied Mathematics* **31**, 119–125.

Gerontidis, I., and R. L. Smith (1982), Monte Carlo generation of order statistics from general distributions, *Applied Statistics* **31**, 238–243.

Geweke, John (1991a), Efficient simulation from the multivariate normal and Student-t distributions subject to linear constraints, *Computer Science and Statistics: Proceedings of the Twenty-third Symposium on the Interface* (edited by Elaine M. Keramidas), Interface Foundation of North America, 571–578.

Geweke, John (1991b), Generic, algorithmic approaches to Monte Carlo integration in Bayesian inference, *Statistical Multiple Integration* (edited by Nancy Flournoy, Robert K. Tsutakawa), American Mathematical Society, 117–135.

Geyer, Charles J. (1992), Practical Markov chain Monte Carlo (with discussion), *Statistical Science* **7**, 473–511.

Gilks, W. R. (1992), Derivative-free adaptive rejection sampling for Gibbs sampling, *Bayesian Statistics 4* (edited by J. M. Bernardo, J. O. Berger, A. P. Dawid, and A. F. M. Smith), Oxford University Press, Oxford, United Kingdom, 641–649.

Gilks, W. R.; N. G. Best; and K. K. C. Tan (1995), Adaptive rejection Metropolis sampling within Gibbs sampling, *Applied Statistics* **44**, 455–472 (Corrections, 1997, *ibid.* **46**, 541–542).

Gilks, W. R.; S. Richardson; and D. J. Spiegelhalter (Editors) (1996), *Markov Chain Monte Carlo in Practice*, Chapman & Hall, London.

Gilks, Walter R., and Gareth O. Roberts (1996), Strategies for improving MCMC, *Practical Markov Chain Monte Carlo* (edited by W. R. Gilks, S. Richardson, and D. J. Spiegelhalter), Chapman & Hall, London, 89–114.

Gilks, W. R.; G. O. Roberts; and E. I. George (1994), Adaptive direction sampling, *The Statistician* **43**, 179–189.

Gilks, W. R.; A. Thomas; and D. J. Spiegelhalter (1992), Software for the Gibbs sampler, *Computing Science and Statistics* **24**, 439–448.

Gilks, W. R., and P. Wild (1992), Adaptive rejection sampling for Gibbs sampling, *Applied Statistics* **41**, 337–348.

Gleser, Leon Jay (1976), A canonical representation for the noncentral Wishart distribution useful for simulation, *Journal of the American Statistical Association* **71**, 690–695.

Golder, E. R., and J. G. Settle (1976), The Box-Muller method for generating pseudo-random normal deviates, *Applied Statistics* **25**, 12–20.

Golomb, S. W. (1982), *Shift Register Sequences*, second edition, Aegean Part Press, Laguna Hills, California.

Gordon, J. (1989), Fast multiplicative inverse in modular arithmetic, *Cryptography and Coding* (edited by H. J. Beker and F. C. Piper), Clarendon Press, Oxford, United Kingdom, 269–279.

Grafton, R. G. T. (1981), The runs-up and runs-down tests, *Applied Statistics* **30**, 81–85.

Greenwood, J. Arthur (1976a), The demands of trivial combinatorial problems on random number generators, *Proceedings of the Ninth Interface Symposium on Computer Science and Statistics* (edited by David Hoaglin and Roy Welsch), Prindle, Weber, and Schmidt, Boston, 222–227.

Greenwood, J. A. (1976b), A fast machine-independent long-period generator for 31-bit pseudo-random numbers, *Compstat 1976: Proceedings in Computational Statistics* (edited by J. Gordesch and P. Naeve), Physica-Verlag, Vienna, 30–36.

Greenwood, J. Arthur (1976c), Moments of time to generate random variables by rejection, *Annals of the Institute for Statistical Mathematics* **28**, 399–401.

Griffiths, P., and I. D. Hill (Editors) (1985), *Applied Statistics Algorithms*, Ellis Horwood Limited, Chichester, United Kingdom.

Gropp, William; E. Lusk; and A. Skjellum (1994), *Using MPI — Portable Parallel Programming with the Message Passing Interface*, The MIT Press, Cambridge, Massachusetts.

Grothe, H. (1987), Matrix generators for pseudo-random vector generation, *Statistische Hefte* **28**, 233–238.

Guerra, Victor O.; Richard A. Tapia; and James R. Thompson (1976), A random number generator for continuous random variables based on an interpolation procedure of Akima, *Computer Science and Statistics: 9th Annual Symposium on the Interface* (edited by David C. Hoaglin and Roy E. Welsch), Prindle, Weber, & Schmidt, Boston, 228–230.

Halton, J. H. (1960), On the efficiency of certain quasi-random sequences of points in evaluating multi-dimensional integrals, *Numerische Mathematik* **2**, 84–90 (Corrections, 1960, *ibid.* **2**, 190).

Hammersley, J. M., and D. C. Handscomb (1964), *Monte Carlo Methods*, Methuen & Co., London.

Hanxleden, Reinhard v., and L. Ridgeway Scott (1992), Correctness and determination of parallel Monte Carlo processes, *Parallel Computing* **18**, 121–132.

Hartley, H. O., and D. L. Harris (1963), Monte Carlo computations in normal correlation procedures, *Journal of the ACM* **10**, 302–306.

Hastings, W. K. (1970), Monte Carlo sampling methods using Markov chains and their applications. *Biometrika* **57**, 97–109.

Heiberger, Richard M. (1978), Generation of random orthogonal matrices, *Applied Statistics* **27**, 199–205. (See Tanner and Thisted, 1982.)

Hellekalek, P. (1984), Regularities of special sequences, *Journal of Number Theory* **18**, 41–55.

Hickernell, Fred J. (1995), A comparison of random and quasirandom points for multidimensional quadrature, *Monte Carlo and Quasi-Monte Carlo Methods in Scientific Computing* (edited by Harald Niederreiter and Peter Jau-Shyong Shiue), Springer-Verlag, New York, 212–227.

Hoaglin, D. C., and D. F. Andrews (1975), The reporting of computation-based results in statistics, *The American Statistician* **29**, 122–126.

Hope, A. C. A. (1968), A simplified Monte Carlo significance test procedure, *Journal of the Royal Statistical Society, Series B* **30**, 582–598.

Hopkins, T. R. (1983), A revised algorithm for the spectral test, *Applied Statistics* **32**, 328–335.

Hörmann, W. (1994a), A universal generator for discrete log-concave distributions, *Computing* **52**, 89–96.

Hörmann, Wolfgang (1994b), A note on the quality of random variates generated by the ratio of uniforms method, *ACM Transactions on Modeling and Computer Simulation* **4**, 96–106.

Hörmann, Wolfgang (1995), A rejection technique for sampling from T-concave distributions, *ACM Transactions on Mathematical Software* **21**, 182–193.

Hörmann, Wolfgang, and Gerhard Derflinger (1993), A portable random number generator well suited for the rejection method, *ACM Transactions on Mathematical Software* **19**, 489–495.

Hörmann, Wolfgang, and Gerhard Derflinger (1994), The transformed rejection method for generating random variables, an alternative to the ratio of uniforms method, *Communications in Statistics — Simulation and Computation* **23**, 847–860.

Huberman, Bernardo A.; Peter L. T. Pirolli; James E. Pitkow; and Rajan M. Lukose (1998), Strong regularities in World Wide Web surfing, *Science* **280**, 95–97.

James, F. (1990), A review of pseudorandom number generators, *Computer Physics Communications* **60**, 329–344.

James, F. (1994), RANLUX: A Fortran implementation of the high-quality pseudorandom number generator of Lüscher, *Computer Physics Communications* **79**, 111–114.

Jöhnk, M. D. (1964), Erzeugung von Betaverteilter und Gammaverteilter Zufallszahlen, *Metrika* **8**, 5–15.

Johnson, Mark E. (1987), *Multivariate Statistical Simulation*, John Wiley & Sons, New York.

Johnson, N. L., and S. Kotz (1987), *Distributions in Statistics. Continuous Multivariate Distributions*, John Wiley & Sons, New York.

Johnson, Valen E. (1996), Studying convergence of Markov chain Monte Carlo algorithms using coupled sample paths, *Journal of the American Statistical Association* **91**, 154–166.

Joy, Corwin; Phelim P. Boyle; and Ken Seng Tan (1996), Quasi-Monte Carlo methods in numerical finance, *Management Science* **42**, 926–938.

Kachitvichyanukul, Voratas (1982), *Computer generation of Poisson, binomial, and hypergeometric random variables*, unpublished Ph.D. dissertation, Purdue University, West Lafayette, Indiana.

Kachitvichyanukul, Voratas; Shiow-Wen Cheng; and Bruce Schmeiser (1988), Fast Poisson and binomial algorithms for correlation induction, *Journal of Statistical Computation and Simulation* **29**, 17–33.

Kachitvichyanukul, Voratas, and Bruce Schmeiser (1985), Computer generation of hypergeometric random variates, *Journal of Statistical Computation and Simulation* **22**, 127–145.

Kachitvichyanukul, Voratas, and Bruce W. Schmeiser (1988), Binomial random variate generation, *Communications of the ACM* **31**, 216–223.

Kachitvichyanukul, Voratas, and Bruce W. Schmeiser (1990), BTPEC: Sampling from the binomial distribution, *ACM Transactions on Mathematical Software* **16**, 394-397.

Kahn, H., and A. W. Marshall (1953), Methods of reducing sample size in Monte Carlo computations, *Journal of Operations Research Society of America* **1**, 263-278.

Kato, Takashi; Li-ming Wu; and Niro Yanagihara (1996a), On a nonlinear congruential pseudorandom number generator, *Mathematics of Computation* **65**, 227–233.

Kato, Takashi; Li-ming Wu; and Niro Yanagihara (1996b), The serial test for a nonlinear pseudorandom number generator, *Mathematics of Computation* **65**, 761–769.

Kemp, A. W. (1981), Efficient generation of logarithmically distributed pseudorandom variables, *Applied Statistics* **30**, 249–253.

Kemp, A. W. (1990), Patchwork rejection algorithms, *Journal of Computational and Applied Mathematics* **31**, 127–131.

Kemp, C. D. (1986), A modal method for generating binomial variables, *Communications in Statistics — Theory and Methods* **15**, 805–813.

Kemp, C. D., and A. W. Kemp (1987), Rapid generation of frequency tables, *Applied Statistics* **36**, 277–282.

Kemp, C. D., and A. W. Kemp (1991), Poisson random variate generation, *Applied Statistics* **40**, 143–158.

Kennedy, William J., and James E. Gentle (1980), *Statistical Computing*, Marcel Dekker, Inc., New York.

Kinderman, A. J., and J. F. Monahan (1977), Computer generation of random variables using the ratio of uniform deviates, *ACM Transaction on Mathematical Software* **3**, 257–260.

Kinderman, A. J., and J. F. Monahan (1980), New methods for generating Student's t and gamma variables, *Computing* **25**, 369–377.

Kinderman, A. J., and J. G. Ramage (1976), Computer generation of normal random variables, *Journal of the American Statistical Association* **71**, 893–896.

Kirkpatrick, Scott, and Erich P. Stoll (1981), A very fast shift-register sequence random number generator, *Journal of Computational Physics* **40**, 517–526.

Kleijnen, Jack P. C. (1977), Robustness of a multiple ranking procedure: A Monte Carlo experiment illustrating design and analysis techniques, *Communications in Statistics — Simulation and Computation* **B6**, 235–262.

Knuth, Donald E. (1975), Estimating the efficiency of backtrack programs, *Mathematics of Computation* **29**, 121–136.

Knuth, Donald E. (1981), *The Art of Computer Programming, Volume 2, Seminumerical Algorithms*, second edition, Addison-Wesley Publishing Company, Reading, Massachusetts.

Kocis, Ladislav, and William J. Whiten (1997), Computational investigations of low-discrepancy sequences, *ACM Transactions on Mathematical Software* **23**, 266–294.

Koehler, J. R., and A. B. Owen (1996), Computer experiments, *Handbook of Statistics, Volume 13* (edited by S. Ghosh and C. R. Rao), Elsevier Science Publishers, Amsterdam, 261–308.

Krawczyk, Hugo (1992), How to predict congruential generators, *Journal of Algorithms* **13** 527–545.

Krommer, Arnold R., and Christoph W. Ueberhuber (1994), *Numerical Integration on Advanced Computer Systems*, Springer-Verlag, New York.

Kronmal, R. A., and A. V. Peterson (1979a), On the alias method for generating random variables from a discrete distribution, *The American Statistician* **33**, 214–218.

Kronmal, R. A., and A. V. Peterson (1979b), The alias and alias-rejection-mixture methods for generating random variables from probability distributions, *Proceedings of the 1979 Winter Simulation Conference*, Institute of Electrical and Electronics Engineers, New York, 269–280.

Kronmal, R. A., and A. V. Peterson (1981), A variant of the acceptance-rejection method for computer generation of random variables, *Journal of the American Statistical Association* **76**, 446–451.

Kronmal, R. A., and A. V. Peterson (1984), An acceptance-complement analogue of the mixture-plus-acceptance-rejection method for generating random variables, *ACM Transactions on Mathematical Software* **10**, 271–281.

Lagarias, Jeffrey C. (1993), Pseudorandom numbers, *Statistical Science* **8**, 31–39.

Laud, Purushottam W.; Paul Ramgopal; and Adrian F. M. Smith (1993), Random variate generation from D-distributions, *Statistics and Computing* **3**, 109–112.

Lawrance, A. J. (1992), Uniformly distributed first-order autoregressive time series models and multiplicative congruential random number generators, *Journal of Applied Probability* **29**, 896–903.

Lawrence, Steve, and C. Lee Giles (1998), Searching the World Wide Web, *Science* **280**, 98–100.

Learmonth, G. P., and P. A. W. Lewis (1973a), *Naval Postgraduate School Random Number Generator Package LLRANDOM, NPS55LW73061A*, Naval Postgraduate School, Monterey, California.

Learmonth, G. P., and P. A. W. Lewis (1973b), Statistical tests of some widely used and recently proposed uniform random number generators, *Computer Science and Statistics: 7th Annual Symposium on the Interface* (edited by William J. Kennedy), Statistical Laboratory, Iowa State University, Ames, Iowa, 163–171.

L'Ecuyer, Pierre (1988), Efficient and portable combined random number generators, *Communications of the ACM* **31**, 742–749, 774.

L'Ecuyer, Pierre (1990), Random numbers for simulation, *Communications of the ACM* **33**, 85–97.

L'Ecuyer, Pierre; François Blouin; and Raymond Couture (1993), A search for good multiple recursive random number generators, *ACM Transactions on Modeling and Computer Simulation* **3**, 87–98.

Lee, A. J. (1993), Generating random binary deviates having fixed marginal distributions and specified degrees of association, *The American Statistician* **47**, 209–215.

Leeb, Hannes, and Stefan Wegenkittl (1997), Inversive and linear congruential pseudorandom number generators in empirical tests, *ACM Transactions on*

Modeling and Computer Simulation **7**, 272–286.

Lehmer, D. H. (1951), Mathematical methods in large-scale computing units, *Proceedings of the Second Symposium on Large Scale Digital Computing Machinery*, Harvard University Press, Cambridge, Massachusetts. 141–146.

Lesk, Michael (1997), *Practical Digital Libraries: Books, Bytes, and Bucks*, Morgan Kaufman Publishers, San Francisco.

Leva, Joseph L. (1992a), A fast normal random number generator, *ACM Transactions on Mathematical Software* **18**, 449–453.

Leva, Joseph L. (1992b), Algorithm 712: A normal random number generator, *ACM Transactions on Mathematical Software* **18**, 454–455.

Lewis, P. A. W.; A. S. Goodman; and J. M. Miller (1969), A pseudo-random number generator for the System/360, *IBM Systems Journal* **8**, 136–146.

Lewis, P. A. W., and E. J. Orav (1989), *Simulation Methodology for Statisticians, Operations Analysts, and Engineers, Volume I*, Wadsworth & Brooks/ Cole, Pacific Grove, California.

Lewis, P. A. W., and G. S. Shedler (1979), Simulation of nonhomogeneous Poisson processes by thinning, *Naval Logistics Quarterly* **26**, 403–413.

Lewis, T. G., and W. H. Payne (1973), Generalized feedback shift register pseudorandom number algorithm, *Journal of the ACM* **20**, 456–468.

Li, Kim-Hung (1994), Reservoir-sampling algorithms of time complexity $O(n(1+ \log(N/n)))$, *ACM Transactions on Mathematical Software* **20**, 481–493.

Li, Shing Ted, and Joseph L. Hammond (1975), Generation of pseudo-random numbers with specified univariate distributions and correlation coefficients, *IEEE Transactions on Systems, Man, and Cybernetics* **5**, 557–560.

Liu, Jun S. (1996), Metropolized independent sampling with comparisons to rejection sampling and importance sampling, *Statistics and Computing* **6**, 113–119.

Luby, M. (1992), Pseudorandom generators from one-way functions, *Lecture Notes in Computer Science*, **576**, 300, Springer-Verlag, New York.

Lurie, D., and H. O. Hartley (1972), Machine generation of order statistics for Monte Carlo computations, *The American Statistician* **26**(1), 26–27.

Lurie, D., and R. L. Mason (1973), Empirical investigation of general techniques for computer generation of order statistics, *Communications in Statistics* **2**, 363–371.

Lurie, Philip M., and Matthew S. Goldberg (1998), An approximate method for for sampling correlated random variables from partially-specified distributions, *Management Science* **44**, 203–218.

Lüscher, Martin (1994), A portable high-quality random number generator for lattice field theory simulations, *Computer Physics Communications* **79**, 100–110.

MacEachern, Steven N., and L. Mark Berliner (1994), Subsampling the Gibbs sampler, *The American Statistician* **48**, 188–190.

MacLaren, M. D., and G. Marsaglia (1965), Uniform random number generators, *Journal of the ACM* **12**, 83–89.

Maclaren, N. M. (1989), The generation of multiple independent sequences of pseudorandom numbers, *Applied Statistics* **38**, 351–359.

Manly, Bryan F. J. (1991), *Randomization and Monte Carlo Methods in Biology*, Chapman & Hall, London.

Marasinghe, Mervyn G., and William J. Kennedy, Jr. (1982), Direct methods for generating extreme characteristic roots of certain random matrices, *Communications in Statistics — Simulation and Computation* **11**, 527–542.

Marriott, F. H. C. (1979), Barnard's Monte Carlo tests: How many simulations?, *Applied Statistics* **28**, 75–78.

Marsaglia, G. (1962), Random variables and computers, *Information Theory, Statistical Decision Functions, and Random Processes* (edited by J. Kozesnik), Czechoslovak Academy of Sciences, Prague, 499–510.

Marsaglia, G. (1963), Generating discrete random variables in a computer, *Communications of the ACM* **6**, 37–38.

Marsaglia, G. (1964), Generating a variable from the tail of a normal distribution, *Technometrics* **6**, 101–102.

Marsaglia, G. (1968), Random numbers fall mainly in the planes, *Proceedings of the National Academy of Sciences* **61**, 25–28.

Marsaglia, G. (1972a), The structure of linear congruential sequences, *Applications of Number Theory to Numerical Analysis* (edited by S. K. Zaremba), Academic Press, New York, 249–286.

Marsaglia, G. (1972b), Choosing a point from the surface of a sphere, *Annals of Mathematical Statistics* **43**, 645–646.

Marsaglia, G. (1977), The squeeze method for generating gamma variates, *Computers and Mathematics with Applications* **3**, 321–325.

Marsaglia, G. (1980), Generating random variables with a *t*-distribution, *Mathematics of Computation* **34**, 235–236.

Marsaglia, George (1984), The exact-approximation method for generating random variables in a computer, *Journal of the American Statistical Association* **79**, 218–221.

Marsaglia, George (1985), A current view of random number generators, *Computer Science and Statistics: 16th Symposium on the Interface* (edited by L. Billard), North-Holland, Amsterdam, 3–10.

Marsaglia, George (1991), Normal (Gaussian) random variables for supercomputers, *Journal of Supercomputing* **5**, 49–55.

Marsaglia, George (1995), *The Marsaglia Random Number CDROM, including the DIEHARD Battery of Tests of Randomness*, Department of Statistics, Florida State University, Tallahassee, Florida.

Marsaglia, G., and T. A. Bray (1964), A convenient method for generating normal variables, *SIAM Review* **6**, 260–264.

Marsaglia, G.; M. D. MacLaren; and T. A. Bray (1964), A fast method for generating normal random variables, *Communications of the ACM* **7**, 4–10.

Marsaglia, George, and Ingram Olkin (1984), Generating correlation matrices, *SIAM Journal on Scientific and Statistical Computing* **5**, 470–475.

Marsaglia, George, and W. W. Tsang (1984), A fast, easily implemented method for sampling from decreasing or symmetric unimodal density functions, *SIAM Journal of Scientific and Statistical Computing* **5**, 349–359.

Marsaglia, George, and Arif Zaman (1991), A new class of random number generators, *The Annals of Applied Probability* **1**, 462–480.

Marsaglia, George; Arif Zaman; and John C. W. Marsaglia (1994), Rapid evaluation of the inverse normal distribution function, *Statistics and Probability Letters* **19**, 259–266.

Mascagni, Michael; Steven A. Cuccaro; Daniel V. Pryor; and M. L. Robinson (1993), Recent developments in parallel pseudorandom number generation, *Proceedings of the Sixth SIAM Conference on Parallel Processing for Scientific Computing* (in two volumes), Society for Industrial and Applied Mathematics, Philadelphia, 524–529.

Mascagni, Michael; M. L. Robinson; Daniel V. Pryor; and Steven A. Cuccaro (1995), Parallel pseudorandom number generation using additive lagged-Fibonacci recursions, *Monte Carlo and Quasi-Monte Carlo Methods in Scientific Computing* (edited by Harald Niederreiter and Peter Jau-Shyong Shiue), Springer-Verlag, New York, 262–267.

Marshall, A. W., and I. Olkin (1967), A multivariate exponential distribution, *Journal of the American Statistical Association* **62**, 30–44.

Matsumoto, Makoto, and Yoshiharu Kurita (1992), Twisted GFSR Generators, *ACM Transactions on Modeling and Computer Simulation* **2**, 179–194.

Matsumoto, Makoto, and Yoshiharu Kurita (1994), Twisted GFSR Generators II, *ACM Transactions on Modeling and Computer Simulation* **4**, 245–266.

McKay, Michael D.; William J. Conover; and Richard J. Beckman (1979), A comparison of three methods for selecting values of input variables in the analysis of output from a computer code, *Technometrics* **21**, 239–245.

McLeod, A. I., and D. R. Bellhouse (1983), A convenient algorithm for drawing a simple random sample, *Applied Statistics* **32**, 182–184.

Mendoza-Blanco, José R., and Xin M. Tu (1997), An algorithm for sampling the degrees of freedom in Bayesian analysis of linear regressions with *t*-distributed errors, *Applied Statistics* **46**, 383–413.

Metropolis, N.; A. W. Rosenbluth; M. N. Rosenbluth; A. H. Teller; E. Teller (1953), Equations of state calculation by fast computing machines, *Journal of Chemical Physics* **21**, 1087–1092.

Meyn, S. P., and R. L. Tweedie (1993), *Markov Chains and Stochastic Stability*, Springer-Verlag, New York.

Mihram, G. A. (1972), *Simulation: Statistical Foundations and Methodology*, Academic Press, New York.

Mihram, George A., and Robert A. Hultquist (1967), A bivariate warning-time/failure-time distribution, *Journal of the American Statistical Association* **62**, 589–599.

Modarres, R., and J. P. Nolan (1994), A method for simulating stable random vectors, *Computational Statistics* **9**, 11–19.

Monahan, John F. (1987), An algorithm for generating chi random variables, *ACM Transactions on Mathematical Software* **13**, 168–171 (Corrections, 1988, *ibid.* **14**, 111).

Morgan, B. J. T. (1984), *Elements of Simulation*, Chapman & Hall, London.

Nagaraja, H. N. (1979), Some relations between order statistics generated by different methods, *Communications in Statistics — Simulation and Computation* **B8**, 369–377.

Neave, H. R. (1973), On using the Box-Muller transformation with multiplicative congruential pseudo-random number generators, *Applied Statistics* **22**, 92–97.

Niederreiter, H. (1988), Remarks on nonlinear congruential pseudorandom numbers, *Metrika* **35**, 321–328.

Niederreiter, H. (1989), The serial test for congruential pseudorandom numbers generated by inversions, *Mathematics of Computation* **52**, 135–144.

Niederreiter, Harald (1992), *Random Number Generation and Quasi-Monte Carlo Methods*, Society for Industrial and Applied Mathematics, Philadelphia.

Niederreiter, Harald (1993), Factorization of polynomials and some linear-algebra problems over finite fields, *Linear Algebra and Its Applications* **192** 301–328.

Niederreiter, Harald (1995a), The multiple-recursive matrix method for pseudoranomd number generation, *Finite Fields and Their Applications* **1**, 3–30.

Niederreiter, Harald (1995b), Pseudorandom vector generation by the multiple-recursive matrix method, *Mathematics of Computation* **64**, 279–294.

Niederreiter, Harald (1995c), New developments in uniform pseudorandom number and vector generation, *Monte Carlo and Quasi-Monte Carlo Methods in Scientific Computing* (edited by Harald Niederreiter and Peter Jau-Shyong Shiue), Springer-Verlag, New York, 87–120.

Niederreiter, Harald (1995d), Some linear and nonlinear methods for pseudorandom number generation, *Proceedings of the 1995 Winter Simulation Conference*, Association for Computing Machinery, New York, 250–254.

Niederreiter, Harald; Peter Hellekalek; Gerhard Larcher; and Peter Zinterhof (Editors) (1997), *Monte Carlo and Quasi-Monte Carlo Methods 1996* Springer-Verlag, New York.

Niederreiter, Harald, and Peter Jau-Shyong Shiue (Editors) (1995), *Monte Carlo and Quasi-Monte Carlo Methods in Scientific Computing* Springer-Verlag, New York.

Norman, J. E., and L. E. Cannon (1972), A computer program for the generation of random variables from any discrete distribution, *Journal of Statistical Computation and Simulation* **1**, 331–348.

Odell, P. L., and A. H. Feiveson (1966), A numerical procedure to generate a sample covariance matrix, *Journal of the American Statistical Association* **61**, 199–203.

Olken, Frank, and Doron Rotem (1995a), Random sampling from databases: A survey, *Statistics and Computing* **5**, 25–42.

Olken, Frank, and Doron Rotem (1995b), Sampling from spatial databases, *Statistics and Computing* **5**, 43–57.

Owen, A. B. (1992a), A central limit theorem for Latin hypercube sampling, *Journal of the Royal Statistical Society, Series B* **54**, 541–551.

Owen, A. B. (1992b), Orthogonal arrays for computer experiments, integration and visualization, *Statistica Sinica* **2**, 439–452.

Owen, A. B. (1994a), Lattice sampling revisited: Monte Carlo variance of means over randomized orthogonal arrays, *Annals of Statistics* **22**, 930–945.

Owen, Art B. (1994b), Controlling correlations in Latin hypercube samples, *Journal of the American Statistical Association* **89**, 1517–1522.

Papageorgiou, A., and J.F. Traub (1996), Beating Monte Carlo, *Risk*, (June), 63–65.

Park, Chul Gyu; Tasung Park; and Dong Wan Shin (1996), A simple method for generating correlated binary variates, *The American Statistician* **50**, 306–310.

Park, Stephen K., and Keith W. Miller (1988), Random number generators: Good ones are hard to find, *Communications of the ACM* **31**, 1192–1201.

Parrish, Rudolph S. (1990), Generating random deviates from multivariate Pearson distributions, *Computational Statistics and Data Analysis* **9**, 283–295.

Patefield, W. M. (1981), An efficient method of generating $r \times c$ tables with given row and column totals, *Applied Statistics* **30**, 91–97.

Perlman, Michael D., and Michael J. Wichura (1975), Sharpening Buffon's needle, *The American Statistician* **29**, 157–163.

Peterson, A. V., and R. A. Kronmal (1982), On mixture methods for the computer generation of random variables, *The American Statistician* **36**, 184–191.

Philippe, Anne (1997), Simulation of right and left truncated gamma distributions by mixtures, *Statistics and Computing* **7**, 173–181.

Propp, J. G., and D. B. Wilson (1996), Exact sampling with coupled Markov chains and applications to statistical mechanics, *Random Structures and Algorithms* **9**, 223–252.

Pullin, D. I. (1979), Generation of normal variates with given sample mean and variance, *Journal of Statistical Computation and Simulation* **9**, 303–309.

Rabinowitz, M., and M. L. Berenson (1974), A comparison of various methods of obtaining random order statistics for Monte-Carlo computations. *The American Statistician* **28**, 27–29.

Rajasekaran, Sanguthevar, and Keith W. Ross (1993), Fast algorithms for generating discrete random variates with changing distributions, *ACM Transactions on Modeling and Computer Simulation* **3**, 1–19.

Ramberg, John S., and Bruce W. Schmeiser (1974), An approximate method for generating asymmetric random variables, *Communications of the ACM* **17**, 78–82.

RAND Corporation (1955), *A Million Random Digits with 100,000 Normal Deviates*, Free Press, Glencoe, Illinois.

Ratnaparkhi, M. V. (1981), Some bivariate distributions of (X, Y) where the conditional distribution of Y, given X, is either beta or unit-gamma, *Statistical Distributions in Scientific Work. Volume 4 – Models, Structures, and Characterizations* (edited by Charles Taillie, Ganapati P. Patil, and Bruno A. Baldessari), D. Reidel Publishing Company, Boston, 389–400.

Reeder, H. A. (1972), Machine generation of order statistics, *The American Statistician* **26**(4), 56–57.

Relles, Daniel A. (1972), A simple algorithm for generating binomial random variables when N is large, *Journal of the American Statistical Association* **67**, 612–613.

Ripley, Brian D. (1987), *Stochastic Simulation*, John Wiley & Sons, New York.

Ripley, Brian D. (1988), Uses and abuses of statistical simulation, *Mathematical Programming* **42**, 53–68.

Robert, Christian P. (1995), Simulation of truncated normal variables, *Statistics and Computing* **5**, 121–125.

Roberts, G. O. (1992), Convergence diagnostics of the Gibbs sampler, *Bayesian Statistics 4* (edited by J. M. Bernardo, J. O. Berger, A. P. Dawid, and A. F. M. Smith), Oxford University Press, Oxford, United Kingdom, 775–782.

Roberts, Gareth O. (1996), Markov chain concepts related to sampling algorithms, *Practical Markov Chain Monte Carlo* (edited by W. R. Gilks, S. Richardson, and D. J. Spiegelhalter), Chapman & Hall, London, 45–57.

Ronning, Gerd (1977), A simple scheme for generating multivariate gamma distributions with non-negative covariance matrix, *Technometrics* **19**, 179–183.

Rosenbaum, Paul R. (1993), Sampling the leaves of a tree with equal probabilities, *Journal of the American Statistical Association* **88**, 1455–1457.

Rosenthal, Jeffrey S. (1995), Minorization conditions and convergence rates for Markov chain Monte Carlo, *Journal of the American Statistical Association* **90**, 558–566.

Rousseeuw, Peter J., and Annick M. Leroy (1987), *Robust Regression and Outlier Detection*, John Wiley & Sons, New York.

Rubin, Donald B. (1987), Comment on Tanner and Wong, "The calculation of posterior distributions by data augmentation", *Journal of the American Statistical Association* **82**, 543–546.

Rubin, Donald B. (1988), Using the SIR algorithm to simulate posterior distributions (with discussion), *Bayesian Statistics 3* (edited by J. M. Bernardo, M. H. DeGroot, D. V. Lindley, and A. F. M. Smith), Oxford University Press, Oxford, United Kingdom, 395–402.

Ryan, T. P. (1980), A new method of generating correlation matrices, *Journal of Statistical Computation and Simulation* **11**, 79–85.

Sacks, Jerome; William J. Welch; Toby J. Mitchell; and Henry P. Wynn (1989), Design and analysis of computer experiments (with discussion), *Statistical Science* **4**, 409–435.

Sarkar, P. K., and M. A. Prasad (1987), A comparative study of pseudo and quasirandom sequences for the solution of integral equations, *Journal of*

Computational Physics **68**, 66–88.

Sarkar, Tapas K. (1996), A composition-alias method for generating gamma variates with shape parameter greater than 1, *ACM Transactions on Mathematical Software* **22**, 484–492.

Särndal, Carl-Erik; Bengt Swensson; and Jan Wretman (1992), *Model Assisted Survey Sampling*, Springer-Verlag, New York.

Schafer, J. L. (1996), *Analysis of Incomplete Multivariate Data by Simulation*, Chapman and Hall, London.

Schervish, Mark J., and Bradley P. Carlin (1992) On the convergence of successive substitution sampling, *Journal of Computational and Graphical Statistics* **1**, 111- 127.

Schmeiser, Bruce (1983), Recent advances in generation of observations from discrete random variates, *Computer Science and Statistics: The Interface* (edited by James E. Gentle), North-Holland Publishing Company, Amsterdam, 154–160.

Schmeiser, Bruce, and A. J. G. Babu (1980), Beta variate generation via exponential majorizing functions, *Operations Research* **28**, 917–926.

Schmeiser, Bruce, and Voratas Kachitvichyanukul (1990), Noninverse correlation induction: Guidelines for algorithm development, *Journal of Computational and Applied Mathematics* **31**, 173–180.

Schmeiser, Bruce, and R. Lal (1980), Squeeze methods for generating gamma variates, *Journal of the American Statistical Association* **75**, 679–682.

Schucany, W. R. (1972), Order statistics in simulation, *Journal of Statistical Computation and Simulation* **1**, 281–286.

Selke, W.; A. L. Talapov; and L. N. Shchur (1993), Cluster-flipping Monte Carlo algorithm and correlations in "good" random number generators, *JETP Letters* **58**, 665–668.

Shaw, J. E. H. (1988), A quasi-random approach to integration in Bayesian statistics, *Annals of Statistics* **16**, 895–914.

Shockley, James E. (1967), *Introduction to Number Theory*, Holt, Rinehart and Winston, Inc. New York.

Sibuya, M. (1961), Exponential and other variable generators, *Annals of the Institute for Statistical Mathematics* **13**, 231–237.

Smith, A. F. M., and G. O. Roberts (1993), Bayesian computation via the Gibbs sampler and related Markov chain Monte Carlo methods, *Journal of the Royal Statistical Society, Series B* **55**, 3–24.

Smith, Robert L. (1984), Efficient Monte Carlo procedures for generating points uniformly distributed over bounded regions, *Operations Research* **32**, 1297–1308.

Smith, W. B., and R. R. Hocking (1972), Wishart variate generator, *Applied Statistics* **21**, 341–345.

Sobol', I. M. (1967), On the distribution of points in a cube and the approximate evaluation of integrals, *USSR Computational Mathematics and Mathematical Physics* **7**, 86–112.

Sobol', I. M. (1976), Uniformly ditributed sequences with an additional uniform property, *USSR Computational Mathematics and Mathematical Physics* **16**, 236–242.

Sowey, E. R. (1972), A chronological and classified bibliography on random number generation and testing, *International Statistical Review*, **40**, 355–371.

Sowey, E. R. (1978), A second classified bibliography on random number generation and testing, *International Statistical Review*, **46**, 89–102.

Sowey, E. R. (1986), A third classified bibliography on random number generation and testing, *Journal of the Royal Statistical Society, Series A* **149**, 83–107.

Spanier, Jerome, and Earl H. Maize (1994), Quasi-random methods for estimating integrals using relatively small samples, *SIAM Review* **36**, 18–44.

Spanier, Jerome, and Keith B. Oldham (1987), *An Atlas of Functions*, Hemisphere Publishing Corporation, Washington. (Also Springer-Verlag, Berlin.)

Stadlober, Ernst (1990), The ratio of uniforms approach for generating discrete random variates, *Journal of Computational and Applied Mathematics* **31**, 181–189.

Stadlober, Ernst (1991), Binomial variate generation: A method based on ratio of uniforms, *The Frontiers of Statistical Computation, Simulation & Modeling* (edited by P. R. Nelson, E. J. Dudewicz, A. Öztürk, and E. C. van der Meulen), American Sciences Press, Columbus, Ohio, 93–112..

Steel, S. J., and N. J. le Roux (1987), A reparameterisation of a bivariate gamma extension, *Communications in Statistics — Theory and Methods* **16**, 293–305.

Stein, Michael (1987), Large sample properties of simulations using Latin hypercube sampling, *Technometrics* **29**, 143–151.

Stewart, G. W. (1980), The efficient generation of random orthogonal matrices with an application to condition estimators, *SIAM Journal of Numerical Analysis* **17**, 403–409.

Stigler, Stephen M. (1978), Mathematical statistics in the early states, *Annals of Statistics* **6**, 239–265.

Stigler, Stephen M. (1991), Stochastic simulation in the nineteenth century. *Statistical Science* **6**, 89–97.

Student (1908a), On the probable error of a mean, *Biometrika* **6**, 1–25.

Student (1908b), Probable error of a correlation coefficient, *Biometrika* **6**, 302–310.

Sullivan, Stephen J. (1993), Another test for randomness, *Communications of the ACM* **33**, Number 7 (July), 108.

Tadikamalla, Pandu R. (1980a), Random sampling from the exponential power distribution, *Journal of the American Statistical Association* **75**, 683–686.

Tadikamalla, Pandu R. (1980b), On simulating non-normal distributions, *Psychometrika* **45**, 273–279.

Tadikamalla, Pandu R., and Norman L. Johnson (1982), Systems of frequency curves generated by transformations of logistic variables, *Biometrika* **69**,

461–465.

Takahasi, K. (1965), Note on the multivariate Burr's distribution, *Annals of the Institute of Statistical Mathematics* **17**, 257–260.

Tang, Boxin (1993), Orthogonal array-based Latin hypercubes, *Journal of the American Statistical Association* **88**, 1392–1397.

Tanner, Martin A. (1996), *Tools for Statistical Inference*, third edition, Springer-Verlag, New York.

Tanner, M. A., and R. A. Thisted (1982), A remark on AS127. Generation of random orthogonal matrices, *Applied Statistics* **31**, 190–192.

Tanner, Martin A., and Wing Hung Wong (1987), The calculation of posterior distributions by data augmentation (with discussion), *Journal of the American Statistical Association* **82**, 528–549.

Tausworthe, R. C. (1965), Random numbers generated by linear recurrence modulo two, *Mathematics of Computation* **19**, 201–209.

Taylor, Malcolm S., and James R. Thompson (1986), Data based random number generation for a multivariate distribution via stochastic simulation, *Computational Statistics & Data Analysis* **4**, 93–101.

Tezuka, Shu (1991), Neave effect also occurs with Tausworthe sequences, *Proceedings of the 1991 Winter Simulation Conference*, Association for Computing Machinery, New York, 1030–1034.

Tezuka, Shu (1993), Polynomial arithmetic analogue of Halton sequences, *ACM Transactions on Modeling and Computer Simulation* **3**, 99–107.

Tezuka, Shu (1995), *Uniform Random Numbers: Theory and Practice*, Kluwer Academic Publishers, Boston.

Tezuka, Shu, and Pierre L'Ecuyer (1992), Analysis of add-with-carry and subtract-with-borrow generators, *Proceedings of the 1992 Winter Simulation Conference*, Association for Computing Machinery, New York, 443–447.

Tezuka, Shu; Pierre L'Ecuyer; and R. Couture (1994), On the lattice structure of the add-with-carry and subtract-with-borrow random number generators, *ACM Transactions on Modeling and Computer Simulation* **3**, 315–331.

Thomas, Andrew; David J. Spiegelhalter; and Wally R. Gilks (1992), BUGS: A program to perform Bayesian inference using Gibbs sampling, *Bayesian Statistics 4* (edited by J. M. Bernardo, J. O. Berger, A. P. Dawid, and A. F. M. Smith), Oxford University Press, Oxford, United Kingdom, 837–842.

Thompson, William J. (1997), *Atlas for Computing Mathematical Functions: An Illustrated Guide for Practitioners with Programs in C and Mathematica*, John Wiley & Sons, New York.

Tierney, Luke (1990), *Lisp-Stat: An Object-Oriented Environment for Statistical Computing and Dynamic Graphics*, John Wiley & Sons, New York.

Tierney, Luke (1991), Exploring posterior distributions using Markov chains, *Computer Science and Statistics: Proceedings of the Twenty-third Symposium on the Interface* (edited by Elaine M. Keramidas), Interface Foundation of North America, 563–570.

Tierney, Luke (1994), Markov chains for exploring posterior distributions (with discussion), *Annals of Statistics* **22**, 1701–1762.

Tierney, Luke (1996), Introduction to general state-space Markov chain theory, *Practical Markov Chain Monte Carlo* (edited by W. R. Gilks, S. Richardson, and D. J. Spiegelhalter), Chapman & Hall, London, 59–74.

Vale, C. David, and Vincent A. Maurelli (1983), Simulating multivariate non-normal distributions, *Pyschometrika* **48**, 465–471.

Vattulainen, I.; T. Ala-Nissila; and K. Kankaala (1994), Physical tests for random numbers in simulations, *Physical Review Letters* **73**, 2513–2516.

Vitter, J. S. (1984), Faster methods for random sampling, *Communications of the ACM* **27**, 703–717.

Vitter, Jeffrey Scott (1985), Random sampling with a reservoir, *ACM Transactions on Mathematical Software* **11**, 37–57..

Von Neumann, J. (1951), *Various Techniques Used in Connection with Random Digits*, NBS Applied Mathematics Series 12, National Bureau of Standards (now National Institute of Standards and Technology), Washington.

Vose, Michael D. (1991), A linear algorithm for generating random numbers with a given distribution, *IEEE Transactions on Software Engineering* **17**, 972–975.

Wakefield, J. C.; A. E. Gelfand; and A. F. M. Smith (1991), Efficient generation of random variates via the ratio-of-uniforms method, *Statistics and Computing* **1**, 129–133.

Walker, A. J. (1977), An efficient method for generating discrete random variables with general distributions, *ACM Transactions on Mathematical Software* **3**, 253–256..

Wallace, C. S. (1976), Transformed rejection generators for gamma and normal pseudo-random variables, *Australian Computer Journal* **8**, 103–105.

Wallace, C. S. (1996), Fast pseudorandom generators for normal and exponential variates, *ACM Transactions on Mathematical Software* **22**, 119–127.

Wichmann, B. A., and I. D. Hill (1982), An efficient and portable pseudo-random number generator, *Applied Statistics* **31**, 188–190 (Corrections, 1984, *ibid.* **33**, 123).

Wilson, David Bruce, and James Gary Propp (1996), How to get an exact sample from a generic Markov chain and sample a random spanning tree from a directed graph, both within the cover time, *Proceedings of the Seventh Annual ACM-SIAM Symposium on Discrete Algorithms*, ACM, New York, 448–457.

Wollan, Peter C. (1992), A portable random number generator for parallel computers, *Communications in Statistics — Simulation and Computation* **21**, 1247–1254.

Wu, Pei-Chi (1997), Multiplicative, congruential random-number generators with multiplier $\pm 2^{k_1} \pm 2^{k_2}$ and modulus $2^p - 1$, *ACM Transactions on Mathematical Software* **23**, 255–265.

Yu, Bin (1995), Comment on Besag et al., "Bayesian computation and stochastic systems": Extracting more diagnostic information from a single run using cusum path plot, *Statistical Science* **10**, 54–58.

Zaremba, S. K. (Editor) (1972), *Applications of Number Theory to Numerical Analysis*, Academic Press, New York.

Zierler, N., and J. Brillhart (1968), On primitive trinomials (mod 2), *Information and Control* **13**, 541–554.

Ziff, Robert M. (1992), Spanning probability in 2D percolation, *Physical Review Letters* **69**, 2670–2673.

Author Index

Subject Index